THE FACTS ON FILE EARTH SCIENCE HANDBOOK

THE FACTS ON FILE EARTH SCIENCE HANDBOOK

THE DIAGRAM GROUP

Checkmark Books®
An imprint of Facts On File, Inc.

The Facts On File Earth Science Handbook

Diagram Visual Information Ltd

Editorial director	Moira Johnston
Editors	Nancy Bailey, Jean Brady, Paul Copperwaite, Eve Daintith, Bridget Giles, Jane Johnson, Reet Nelis, Jamie Stokes
Design	Richard Hummerstone, Edward Kinsey
Design production	Carole Dease, Oscar Lobban, Lee Lawrence
Artists	Susan Kinsey, Lee Lawrence, Kathleen McDougal
Research	Peter Dease, Catherine & Neil McKenna,
Contributors	Michael Allaby, Martyn Bramwell, John Daintith, Trevor Day, John Haywood, Jim Henderson, David Lambert, Catherine Riches, Dr Robert Youngson
Indexer	Christine Ivamy

Checkmark Books
An imprint of Facts On File, Inc.
11 Penn Plaza
New York, NY 10001

Library of Congress Cataloging-in-Publication Data

The Facts on File earth science handbook / The Diagram Group.
p. cm.
Includes index
ISBN 0-8160-4081-8 (hc) (acid-free paper)—ISBN 0-8160-4586-0 (pbk)
I. Earth sciences—Handbooks, manuals, etc. I. Diagram Group.

QE5 .F32 2000
550—dc21

99-048564

Checkmark Books are available at special discounts when purchased in bulk quantities for businesses, associations, institutions, or sales promotions. Please call our Special Sales Department in New York at 212/967-8800 or 800/322-8755.

You can find Facts On File on the World Wide Web at http://www.factsonfile.com

Cover design by Cathy Rincon

Printed in the United States of America

MP DIAG 10 9 8 7 6 5 4 3 2
(pbk) 10 9 8 7 6 5 4 3 2 1

This book is printed on acid-free paper.

INTRODUCTION

An understanding of science is the basis of all technological advances. Our domestic lives, possessions, cities, and industries have only been developed through scientific research into the principles that underpin the physical world. But obtaining a full view of any branch of science may be difficult without resorting to a range of books. Dictionaries of terms, encyclopedias of facts, biographical dictionaries, chronologies of scientific events – all these collections of facts usually encompass a range of science subjects. THE FACTS ON FILE HANDBOOK LIBRARY covers four major scientific areas – CHEMISTRY, PHYSICS, EARTH SCIENCE (including astronomy), and BIOLOGY.

THE FACTS ON FILE EARTH SCIENCE HANDBOOK has four sections – a glossary of terms, biographies of personalities, a chronology of events, essential charts and tables, and finally an index.

GLOSSARY
The specialized words used in any science subject mean that students need a glossary in order to understand the processes involved. THE FACTS ON FILE EARTH SCIENCE HANDBOOK glossary contains more than 1,400 entries, often accompanied by labeled diagrams to help clarify the meanings.

BIOGRAPHIES
The giants of science – Darwin, Galileo, Einstein, Marie Curie – are widely known, but many other dedicated scientists have also contributed to the advancement of scientific knowledge. THE FACTS ON FILE EARTH SCIENCE HANDBOOK contains biographies of more than 250 people, many of whose achievements may have gone unnoticed but whose discoveries have pushed forward the world's understanding of earth science.

CHRONOLOGY
Scientific discoveries often have no immediate impact. Nevertheless, their effects can influence our lives more than wars, political changes, and world rulers. THE FACTS ON FILE EARTH SCIENCE HANDBOOK covers nearly 3,700 years of events in the history of discoveries in earth science.

CHARTS & TABLES
Basic information on any subject can be hard to find, and books tend to be descriptive. THE FACTS ON FILE EARTH SCIENCE HANDBOOK puts together key charts and tables for easy reference. Scientific discoveries mean that any compilation of facts can never be comprehensive. Nevertheless, this assembly of current information on the subject offers an important resource for today's students.

In past centuries scientists were curious about a wide range of sciences. Today, with disciplines being so independent, students of one subject rarely learn much about the others. THE FACTS ON FILE HANDBOOKS enable students to compare knowledge in chemistry, physics, earth science, and biology, to put each subject in context, and to underline the close connections between all the sciences.

CONTENTS

SECTION ONE GLOSSARY

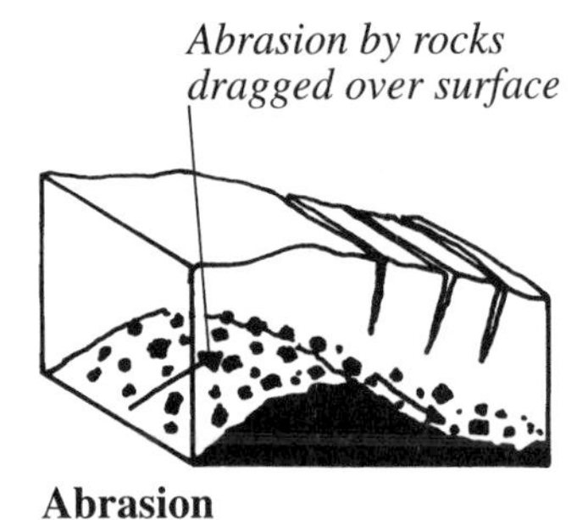

Abrasion

abiogenesis The discredited notion that life commonly arises from nonliving sources, as in the belief that maggots are generated by putrefying meat. Also known as spontaneous generation, this process was formerly believed to explain the origin of microorganisms.

ablation (1) Removal of rock material by weathering, especially wind action.
(2) The removal of snow and ice by melting or sublimation. *See also* erosion.

abrasion Wearing away and remodeling of bedrock by rock and other particles moved by wind, waves, running water, or glacier movement. *See also* erosion.

absolute age The length of time since a particular geological event, expressed in years, as distinct from the age relative to other events. The term does not imply precision.

absolute plate motion The movement of a tectonic plate in the lithosphere relative to some fixed reference such as an area of high volcanic activity (hot spot) or the paleomagnetic poles.

abyssal hills Low-altitude hills on the ocean floor, mainly below 750 feet (250 m) high and a few kilometers in width. They are characteristic of the Pacific floor.

abyssal plain A nearly level tract of the deep ocean floor, about 2–4 miles (3–6.5 km) below the surface of the sea.

abyssal zone The parts of the oceans at depths of about 6,500 feet (2,000 m) or more to which virtually no light penetrates. Organisms living in the abyssal zone have adapted to conditions of darkness, near-zero temperatures, and great pressure.

accessory mineral Mineral present in a rock in such a small quantity that it can be ignored in deciding on the classification of the rock.

accretion The process by which a rock or any other inorganic body grows in size by the addition of particles to its exterior. The term is also applied to the accumulation of a sediment and to the enlargement of a continent.

accretionary wedge A thickened wedge of sediment occurring on the landward side of some oceanic trenches and consisting of sediment scraped off the subducting plate.

accumulation zone The part of a glacier at which the average annual gain in mass from ice, firn, and snow, is greater than the average annual loss.

acid rain Rain acidified by sulfuric, nitric, and other acids that form when water and sunlight react with sulfur dioxide, nitrogen oxides, and other pollutants released by burning fossil fuels. Acid rain can poison lakes, kill forests, and corrode buildings.

1

2

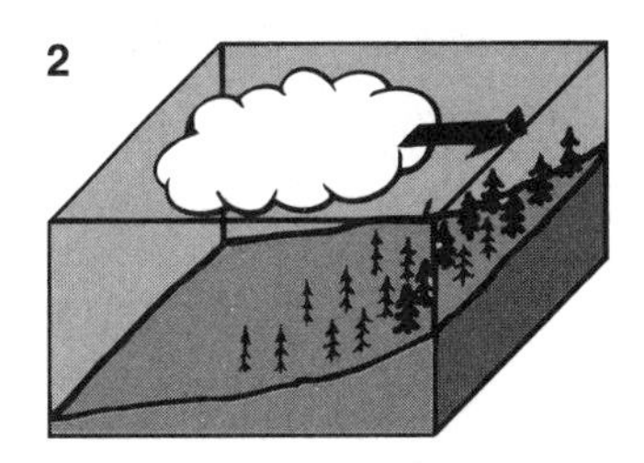

3

4

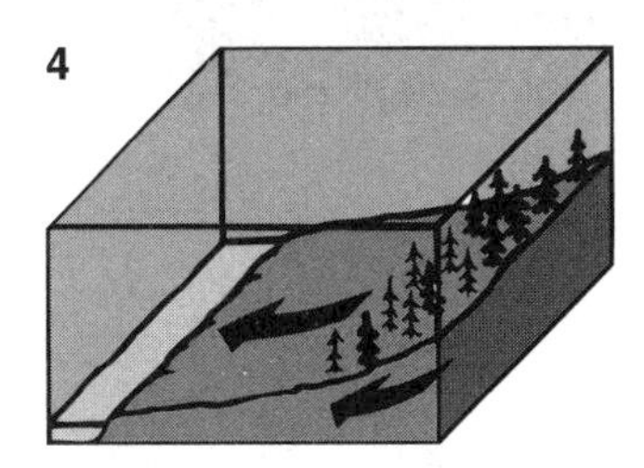

Acid rain

acid rocks Igneous rocks containing more than 63% quartz, or rocks with more than 10% visible quartz.

acid soil Soil with a pH of less than 7. Very acid soil has a pH of less than 5.

actinic radiation Radiation that produces a photochemical effect, especially light and ultraviolet radiation. The Sun is a major source of actinic electromagnetic radiation.

actinium series A series of decaying radioactive elements starting with uranium-235 and ending in an isotope of lead. Each member of the series derives from the radioactive decay of its predecessor.

active layer In regions of permanently frozen ground known as permafrost, this is the upper layer of soil that thaws in summer while the soil below remains frozen.

adamantine luster Having the luster of a polished diamond.

adamellite granites Granites containing alkali and plagioclase feldspars in roughly equal quantities.

adaptive radiation or **cladogenesis** A rapid phase of evolution in which a variety of species develop from a single ancestral form in response to a number of different habitats.

adiabatic process Changes in temperature, pressure, and volume in a quantity of air that involve no exchange of energy or mass with the surrounding atmosphere.

adit A tunnel driven horizontally into a slope for mining minerals or coal, found as veins or seams in a mountain or hill.

adobe A fine, silty, often calcarous sedimented clay occurring in dried desert lake basins. The source is eroded, wind-blown loess.

aeolian *See* eolian.

aerogenerator A windmill that converts the wind's mechanical energy into electricity. *See* wind power.

afforestation Large-scale tree planting. Afforestation provides useful wood and tree roots that grip the soil, protecting steep slopes from erosion.

African plate One of the largest of the present lithospheric plates consisting of the whole of continental Africa.

aftershock An earth tremor that follows a major earthquake. Aftershocks may be multiple and often compound the damage done by the primary shock. The amplitude of an aftershock may approach that of the original shock.

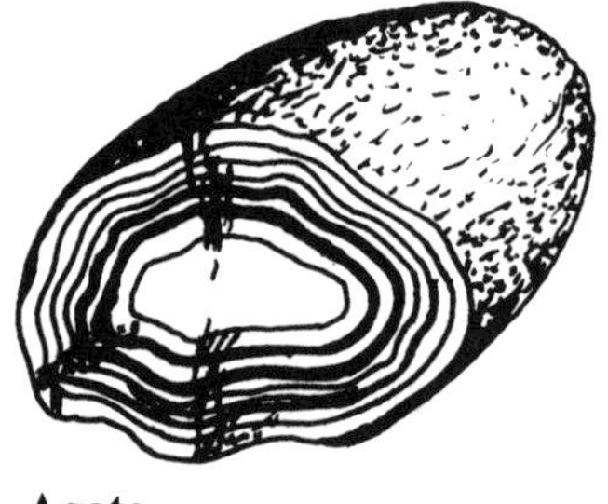

Agate

agate A very finely crystalline form of quartz, often with iron and manganese conferring a decorative banding appearance. Fine specimens are regarded as semiprecious stones.

age A geologic time unit that is a subdivision of an epoch. An age is the smallest unit of time normally employed in geology.

agglomerate A pyroclastic rock formed when volcanic detritus consolidates. It features many rounded fragments greater than 1.23 inches (32 mm) in diameter embedded in a fine-grained matrix of tuff.

agglutinate A pyroclastic rock formed mainly from initially semi-solid clots of liquid lava, thrown out from the vent of a volcano, to which other similar clots adhere during solidification. Also applied to aggregates formed by the impact of micrometeorites on lunar soil.

aggradation *See* alluviation.

agrochemicals The chemical fertilizers and pesticides underpinning modern agriculture.

agrometeorology The study of the relationship between the lower layers of the atmosphere and the Earth's surface, with special emphasis on the effects of this relationship to agriculture.

Agulhas current A surface-water current flowing southwesterly off the east coast of southern Africa, part of the circulation of the southern Indian Ocean.

air The mixture of gases forming the Earth's atmosphere. Air is about 78% nitrogen, 21% oxygen, and 1% other gases, including carbon dioxide and the noble gases.

air mass A meteorological term for an extensive mass of air with broadly uniform temperature and humidity. Depending on its place of origin, it can be tropical or polar, and maritime or continental. Major air masses may be millions of square miles in horizontal extent. Vertically, an air mass may be homogeneous in terms of heat and moisture as a result of thorough mixing, or may preserve a high degree of stratification. The vertical distribution of water vapor and heat determine the primary weather characteristics of an air mass.

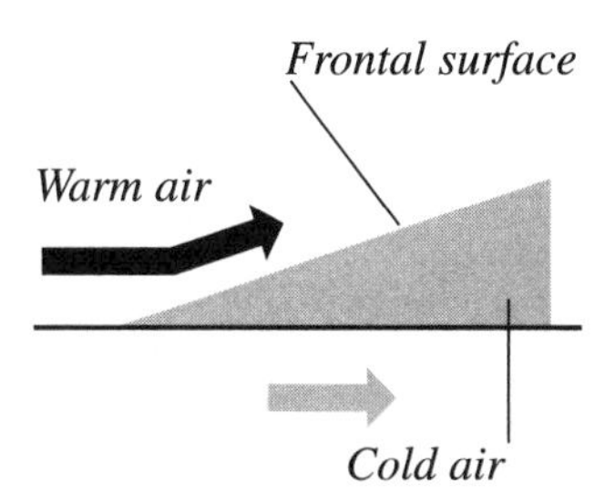

Air mass (warm front)

air mass analysis or **frontal analysis** The analysis of the prevailing air masses on a weather map and of the transition zones and fronts that separate them. At the fronts between air masses, the colder air mass usually passes under the warmer mass like a wedge, resulting in a slope at the interface.

air pollution Contamination of the air, especially by smoke or gases from vehicles, factories, and power stations. It can cause diseases, kill plants, and damage structures.

albedo The proportion of incoming sunlight reflected by the Earth, the atmospheric clouds, and the atmosphere, without causing heating of the Earth's surface. The Earth reflects an average of about 30% of the light it receives from the Sun, but this covers a range from about 8% from some grasses to 90% for fresh snow. Thick cumulus clouds can reflect 80% of the light.

albic Of a very white soil with little or no clay or oxides on the sand particles.

Aleutian trench The oceanic trench marking the boundary between the Pacific lithospheric plate and the North American plate.

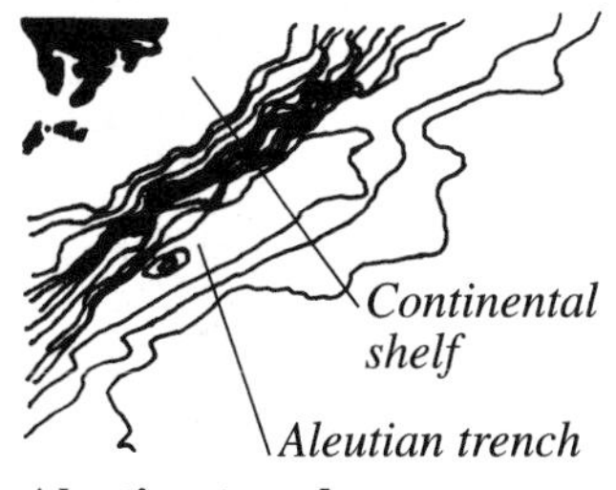

Aleutian trench

algal bloom The sudden appearance of a relative overgrowth of alga in an aquatic ecosystem due to pollution that provides abnormal enrichment for the growth of algae, or to a natural excess of growth over consumption by fish and other water herbivores.

alkali basalt A fine-grained, dark-colored igneous rock containing phenocrysts of olivine, augite, iron oxides, and plagioclase and a higher proportion of sodium and potassium oxides relative to quartz than other basalts.

alkali granites Granites containing only alkali feldspar (orthoclase) or a very high percentage of alkali feldspar (over 66%) and a low percentage of plagioclase feldspar.

alkali metals The elements in the first group of the periodic table, specifically, lithium, sodium, potassium, rubidium, cesium, and francium. All are highly reactive metals that, with the exception of lithium, react violently with water to form hydroxides.

alkaline Describing igneous rocks with a high content of the alkali metal oxides.

alkaline earth metals The elements in the second group of the periodic table, specifically, beryllium, magnesium, calcium, strontium, and barium. The designation "earth" refers to their oxides. All react with water to form hydroxides, and, on heating, these form the oxide and water.

alkaline igneous rock Rock containing a ratio of alkaline oxides (of potasium, sodium, lithium, caesium, etc.) to silica that is higher than 1:6.

Alké pattern The pattern of dunes formed in areas of large quantities of fine sand when exposed to wind of a constantly prevailing direction. The pattern consists of curved segments that are alternately concave to the wind and convex to it.

allochthon A rock that, since the time of its formation, has been moved to a different site as a result of forces acting on the Earth's crust. *Compare* autochthon.

allotriomorphic *See* anhedral.

algal bloom – allotriomorphic

alluvial fan A fan-shaped layer of alluvium deposited by a mountain stream flowing through a narrow valley onto a plain.

Alluvial fan

alluviation or **aggradation** The deposit of material carried in water.

alluvium Silt, sand, or gravel deposited by moving water, either along the bed of a stream or on land that the water overflows. Alluvium builds alluvial fans, deltas, and the floors of floodplains.

alpine Resembling or relating to the European Alps: mid-latitude mountains with sharp peaks sculpted by frost and ice.

altocumulus Fleecy, mid-altitude, fine-weather clouds often forming a broken, banded layer.

altostratus A gray or white sheet of mid-altitude cloud covering the sky.

aluminum A silvery lightweight metal: the most plentiful metallic element and third most plentiful of all elements in the Earth's crust. Feldspars, the largest class of minerals, are aluminum silicates, mainly of potassium, sodium, and calcium.

amber Fossil resin. Amber occasionally contains preserved insects, some of which are extinct.

amethyst Crystalline quartz of gemstone quality with a purplish color derived from traces of ferric iron.

ammonite *See* ammonoid.

ammonoid or **ammonite** An extinct cephalopod animal similar to the contemporary nautilus that is found in large numbers as characteristic fossils, often known as ammonites. They were very abundant during the Mesozoic era.

amphibians Organisms capable of living both on land and in water. Amphibians evolved about 370 million years ago in the Devonian period and were the first vertebrates to live on land.

amphibole Any of a group of rock-forming ferromagnesian silicates.

amphidromic point A nodal point in a sea at which the tidal range is zero and around which the peak of a standing wave or a high-water level rotates once in each tidal cycle.

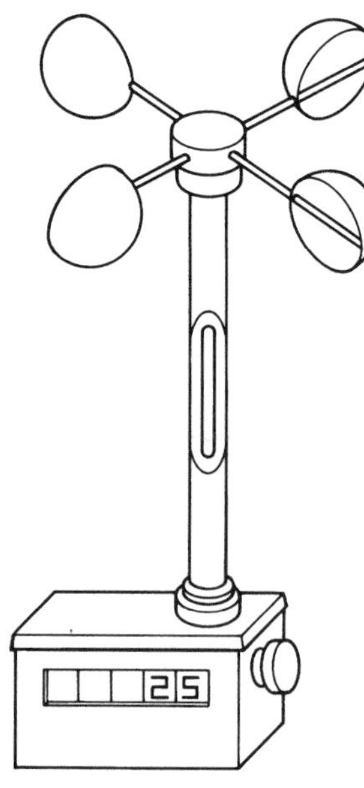

Anemometer

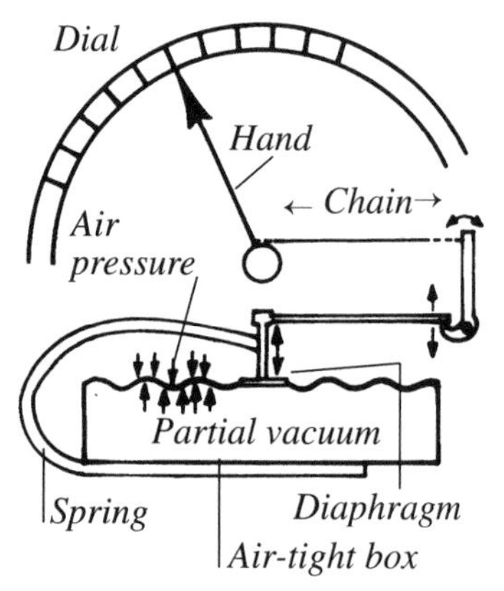

Aneroid barometer

amygdale A gas- or steam-formed cavity in rock that is wholly or partly filled with one or more other minerals.

andesite A dark, fine-grained type of extrusive igneous rock. It takes its name from the Andes Mountains, one of the regions where it occurs.

anemometer An instrument for measuring wind speed. The cup anemometer, the commonest type, indicates wind speed by the rate at which the wind makes a set of cups on horizontal arms spin around a vertical axis.

aneroid barometer An instrument for measuring atmospheric pressure. It features a near vacuum in a container with flexible sides that move in and out as air pressure changes. One end of this bellows-like structure is fixed; the other is connected to a system of levers so that small movements are magnified and move a needle over a scale calibrated in pressure units.

anhedral or **allotriomorphic** Of grains in igneous rock that show no development of regular crystalline facets. This occurs when the free growth of crystals in a melt is interfered with by the presence of already formed crystals.

anhydrite The soft, colorless, white, or grayish mineral anhydrous calcium sulfate. Anhydrite may occur as a cap rock above a salt dome. It has commercial value as an ingredient in cement.

anhydrous Free from water.

Animalia One of the five taxonomic kingdoms of living things, the others being Fungi, Monera, Protista, and Plantae. By definition, an animal cannot generate its own necessary food and must therefore depend on plants or other animals for its nutrition.

anisotropic Having physical properties of unequal value in different crystallographic directions. Applied mainly to the optical properties, such as refractive index, of crystals.

anorthosite A type of intrusive igneous rock consisting of 90% or more of plagioclase feldspar. Anorthosite occurs mainly in Precambrian rocks and on the Moon.

Antarctic (1) Regions south of the Antarctic Circle.
(2) Regions of or around the South Pole with a very cold climate.

Antarctic Circle A line of latitude 66° 30' south of the Equator. It is the Southern Hemisphere's equivalent of the Arctic Circle.

antecedent stream A waterway that has been able to maintain the continuity of its original course in spite of an uplifting of the land to form a ridge roughly at right angles to the course. Because of slow uprising of the land, the waterway has had time to cut a gorge.

anthracite A hard, shiny type of coal with a very high carbon content. Anthracite forms from coal subjected to high pressures and heat deep underground.

anticline A broad arch-like upfold in layered rocks, caused by their compression. In an anticline, the oldest rocks are on the inside of the arch. The opposite of a syncline.

anticlinorium A region, at least several miles across, containing numerous anticlines and synclines.

anticyclone A large mass of high-pressure air bringing calm weather. In temperate regions, this can be hot and sunny in summer, and cold and clear or foggy in winter. An anticyclone is a source region for an air mass and is often slow-moving or stationary. Winds blow clockwise around an anticyclone in the Northern Hemisphere, counterclockwise in the Southern Hemisphere.

apatite The most abundant and widely distributed mineral of the phosphate group. A usually green or gray-green fluorophosphate or chlorphosphate of calcium, with a hardness of 5, that occurs as an accessory mineral in igneous rock and pegmatites. Apatite is used as a phosphate fertilizer.

aphelion The point in a planet's orbit farthest from the Sun.

aphotic Without light.

aphotic zone That part of the open-ocean environment that receives insufficient sunlight to allow photosynthesis to occur. The zone exists at all depths below about 160–320 feet (50–100 m) varying with the latitude and seasons.

Appalachian orogenic belt A belt extending from Newfoundland to Alabama that has undergone compressional tectonics, resulting in an upthrust land form.

aquamarine Beryl.

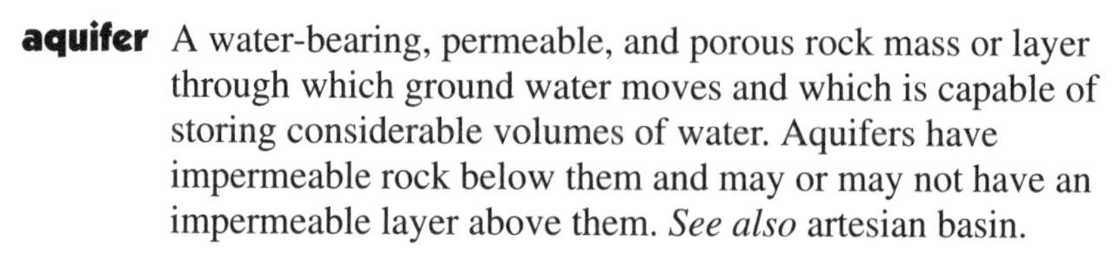

aquifer A water-bearing, permeable, and porous rock mass or layer through which ground water moves and which is capable of storing considerable volumes of water. Aquifers have impermeable rock below them and may or may not have an impermeable layer above them. *See also* artesian basin.

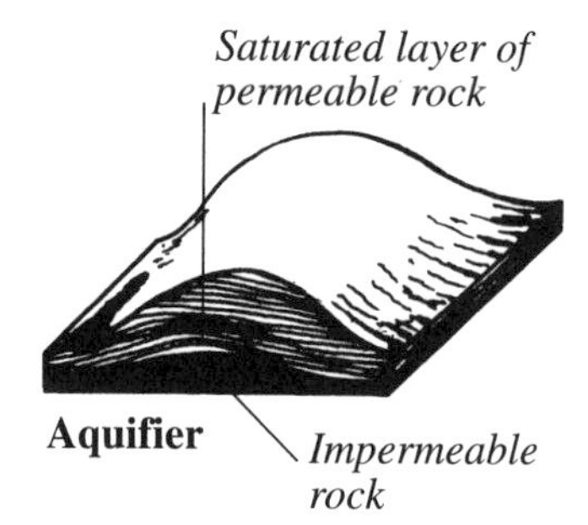

Aquifer

Arabian Plate A contemporary minor lithospheric plate that is separated from the African plate by the widening Red Sea, the Dead Sea system, and the Gulf of Aden.

aragonite A plentiful form of calcium carbonate that is harder than calcite. It is colorless or whitish, gray, or yellowish, and occurs in seabed muds, hot springs, stalactites and stalagmites, veins and cavities, and various rocks.

Archean *See* Archeozoic era.

Archeocetes A suborder of extinct whales, or Cetacea.

Archeocyatha A phylum of extinct animals that functioned in the manner of corals to construct reefs.

Archeogastropoda An order of gastropods, most of which are extinct, that appeared in the Lower Cambrian but that includes the contemporary limpet *Patella vulgata*.

archeomagnetism The study of the magnetic properties of rocks and other materials dating from the remote past. The discipline includes magnetic dating and the use of magnetic-domain orientation to show geologic changes in gross orientation.

archeopteryx An extinct creature with reptilian features that is regarded as the first known bird. Five specimens of its fossils have been found in strata more than 140 million years old.

Archeozoic era or **Archean** The oldest geological era (4–2.5 billion years ago), in which there was only the most primitive and undeveloped forms of life. The earliest of the three divisions of the Precambrian.

Arctic (1) Regions north of the Arctic Circle.
(2) Regions of or around the North Pole with a very cold climate.

Arctic Circle A line of latitude 66° 30' north of the Equator: the Northern Hemisphere's equivalent of the Antarctic Circle. North of the Arctic Circle the Sun does not rise on at least one day in winter or set on at least one day in summer.

arenaceous rocks Sedimentary rocks known also as sandstones. They consist largely of sand particles with a medium grain size, larger than that found in clays.

arenite Sedimentary rock, made of sand-sized particles, in which less than 15% is mud matrix. Arenite grains may be quartz, feldspar, or fragments of other rocks.

arête A mountain ridge with a knifelike crest and very steep flanks. It occurs where the headwalls of two cirques erode until they meet back to back. *See* horn.

argillaceous rocks Fine-grained sedimentary rocks such as clay, mudstone, and shale. They consist largely of tiny clay particles.

arid Lacking enough rainfall for most kinds of vegetation to grow.

arkose Coarse sandstone consisting mainly of quartz grains, but with 25% or more of feldspar grains, held together by a natural mud cement matrix.

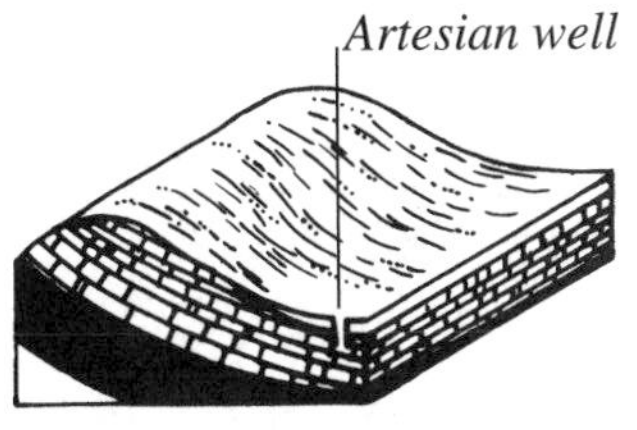

Artesian basin

artesian basin Saucer-like sedimentary rock layers embedded so that impermeable layers sandwich an aquifer. Water may spurt from an artesian well drilled down to the aquifer.

artesian well A well whose water is under high hydrostatic pressure so that it is forced naturally upwards. This arises because the sources of the well are in strata above the level of the well, which is in impermeable rock and the outlet of the well has penetrated this rock.

arthropod A member of the phylum Arthropoda, which is the largest subdivision of the animal kingdom. They are invertebrates with segmented bodies and jointed appendages, mainly legs, and include all the insects, scorpions, crabs, centipedes, and the fossils known as trilobites.

ash Fine-grained lava particles emitted by volcanic eruptions. Light, windborne ash can sometimes travel thousands of miles.

ash-flow tufts Rocks containing fragments picked up from the ground by the pyroclastic flow from a volcano before it hardens. Ash-flow tufts are often found in extensive flat sheets and display a distinctive streaky appearance. Depending on the conditions at the time of formation, the content may be flattened together or welded.

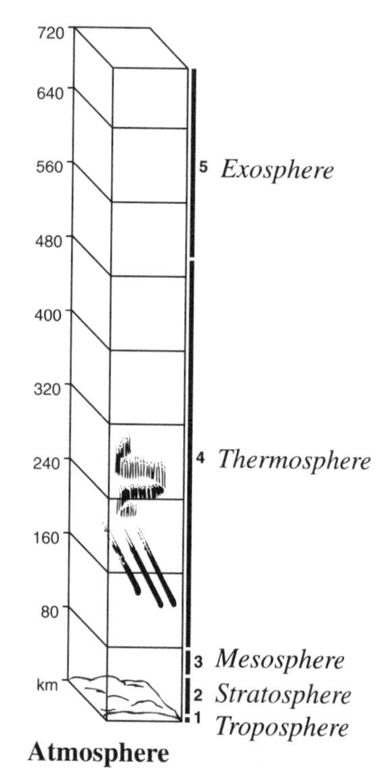

Atmosphere

asphalt lake A pool of bitumen (a mixture of solid or semisolid hydrocarbons) occurring mainly in the southern states of North America and in South America.

asteroids Metallic and stony lumps of rock orbiting the Sun in their thousands.

asthenosphere A dense, plastic, semi-molten layer of mantle just below the lithosphere.

astronomical unit (AU) The average distance between the Sun and the Earth.

atmosphere The envelope of gases surrounding a planet and held there by its gravitational force of attraction. Apart from providing an essential supply of oxygen, the atmosphere controls Earth surface temperature and shields against the harmful effects of solar radiation, making life possible. Changing atmospheric conditions – movement (wind), temperature, water content, and water precipitation – constitute weather.

atmospheric composition The chemical makeup of the atmosphere. There are three major components – nitrogen, oxygen, and argon – and a large number of minor components, including carbon dioxide, helium, neon, krypton, xenon, hydrogen, and a range of oxides of nitrogen. Many other components are present in trace quantities as a result of volcanic and industrial activities. The chemical content of the atmosphere has varied considerably, even radically, since the formation of the Earth.

atmospheric layers The layer closest to the Earth, and the region of the weather, is called the troposphere. Above this is the stratosphere, which extends from about 6–10 miles to about 31 miles (10–16 km to 50 km). The mesosphere extends above the stratosphere from 34 miles (55 km) to about 50 miles (80 km), and contains the mesopause and most of the D layer, a region of low ionization. Above this is the thermosphere, which extends from 50 miles (80 km) to the edge of the atmosphere. This layer receives powerful radiant energy directly from the Sun and shows phenomena such as the aurora. Finally, the highest part of the atmosphere, the exosphere, has very few molecules with appropriate velocities to escape into outer space.

atmospheric pressure Pressure exerted by a planet's atmosphere because of its weight. Atmospheric pressure and density both decrease rapidly, and roughly exponentially, with altitude. Atmospheric pressure is low in cyclones and high in anticyclones.

atmospheric radiation Infrared radiation emitted or transmitted by the atmosphere.

atmospheric window The narrow band of wavelengths of infrared radiation from the Earth that is only minimally absorbed by water vapor in the atmosphere and that may escape into space unless absorbed by clouds.

atoll A ring-shaped coral reef enclosing a lagoon. Atolls form in the tropics on the subsiding rims of collapsed volcanoes.

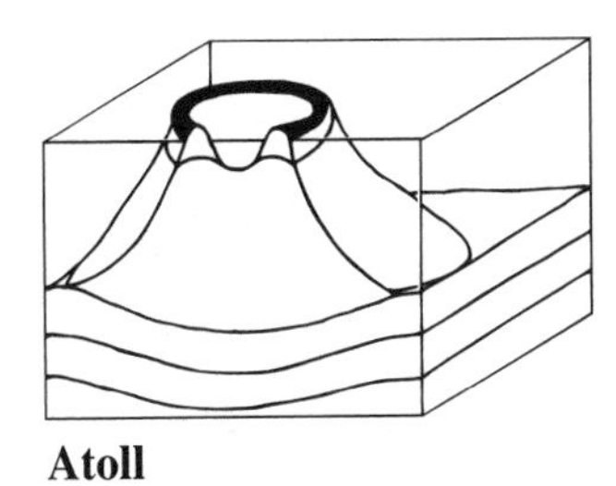
Atoll

atom The invisible, tiny basic unit of matter in a chemical element. Atoms themselves consist of still smaller subatomic particles. *See also* proton, neutron, electron.

atomic energy *See* nuclear energy.

augen A lens-like or eye-like concentration of feldspar or quartz crystals often found in gneisses.

aureole The zone around an igneous intrusion in which the local rocks have suffered metamorphism through heat.

aurora An electrical discharge producing curtains of light, seen in higher latitudes in the night sky.

authigenesis The crystallization of minerals within a sediment during or after its deposition.

autochthon A large body of rock that has remained at the site of its formation.

automatic weather station A collection of meteorological measuring and recording instruments capable, without human intervention, of transmitting data on temperature, pressure, humidity, wind speed and direction, etc. to a central point.

autumn or **fall** In temperate zones, the season between summer and winter.

avalanche A great mass of snow or rock that suddenly slides down a mountainside. Also known as mass-wasting.

avalanche wind A mass of high-speed and often destructive wind occurring ahead of an avalanche.

Aves The class of all birds, from archeopteryx to those of the present.

avulsion The sideways displacement of a stream from its main channel into a new course, occurring in its floodplain.

axis The Earth's axis is an imaginary line linking the North and South Poles through the middle of the Earth. The spinning Earth revolves once around its axis every 24 hours.

axis of rotation The line around which a rotating body rotates. An axis need not be immobile. The Earth's axis undergoes a regular movement, as on the surface of a cone, known as precession.

azimuth (1) The horizontal clockwise angle between a bearing on the Earth's surface and a standard direction, usually north.
(2) In astronomy, the angular clockwise distance from the

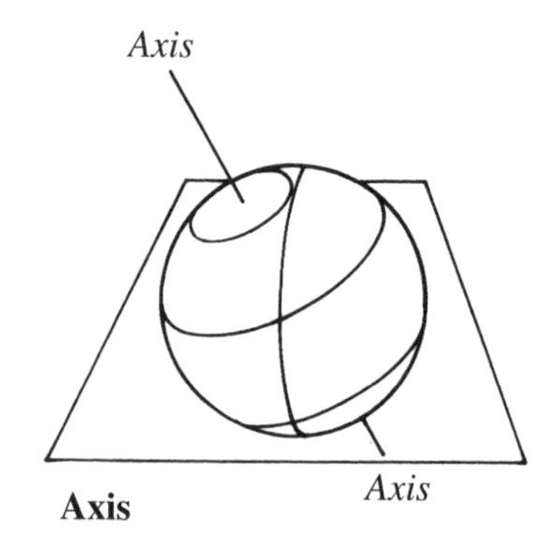

Axis

south point on the horizon to the intersection with the horizon of the vertical circle passing through a star.
(3) In navigation, as in (2) but measured from the north point on the horizon.

azimuthal projections or **zenithal projections** Map projections made by projecting a globe's surface onto a flat sheet touching the globe at a selected point. Azimuthal projections can accurately show relative areas or distances.

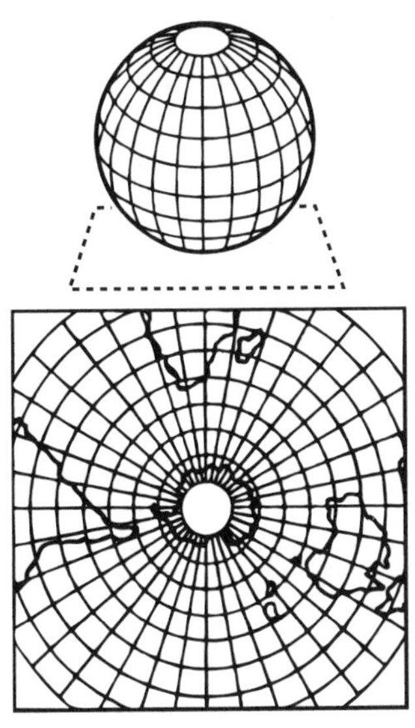

Azimuthal projection

background radiation The low-intensity radiation from natural and human-made sources that is constantly present in the environment. This radiation comes from radioactive substances in rocks and soil and radiation from outer space. Microwave background radiation derives from the original big bang.

backwash Water flowing back to the sea from a wave that has broken as it surges up a beach.

badlands Arid, barren, hilly land, deeply eroded by gullies. South Dakota has some of North America's best-known examples.

bar A muddy, sandy, or shingly ridge lying across a river mouth or a bay.

barchan A sand dune shaped like a crescent, with its points blown forward by the wind. Barchans more than 100 feet (30 m) high occur in the Sahara Desert and Central Asian deserts.

barometer An instrument for measuring atmospheric pressure. In a mercury barometer, pressure supports a column of mercury in a glass tube. *See also* aneroid barometer.

barometric pressure *See* atmospheric pressure.

barrage A fixed or movable wall placed across a watercourse to hold back water, for irrigation or flood control or to generate electricity.

barrier beach An elongated ridge of sand and gravel lying roughly parallel to a coastline but separated from it by a lagoon or a tidal marsh.

barrier island A long, low, narrow, sandy island separated from the mainland by a lagoon.

barrier reef A broad, offshore coral reef aligned with a coast. Between the reef and the coast lies a broad, deep lagoon. The Great Barrier Reef off Queensland, Australia, is the largest on Earth.

basal conglomerate A coarse, clastic sedimentary rock formation made of fragments of roughly equal size, forming the base of a sedimentary sequence resting on an eroded surface.

basalt A dark-colored, fine-grained type of extrusive, basic (as opposed to acid) igneous rock. Most volcanic rock is basalt, which is composed of plagioclase feldspar, pyroxine, magnetite, and often olivine. Basalt flows cover about 70% of the Earth's surface.

base level The limit to which continental erosion can continue. The ultimate base level is sea level, but local base levels may occur at any height above sea level at basins or areas containing rock highly resistant to erosion.

base metal Any of the common metals, such as tin, copper, and lead, as distinct from the precious metals, such as platinum, gold, and silver.

basic Of igneous rocks that contain olivine, pyroxines, and calcium-rich feldspars but no quartz.

basin (1) An area in which the rock has a natural concavity toward a lowest point. Geological basins may be hundreds of miles wide and can determine geological features such as sedimentary rock.
(2) An area drained by a river and its tributaries.

batholith A large dome-shaped, deep-seated, intrusive body of igneous rock of which the lower extremity cannot be determined. Batholiths are often of granite.

bathyal zone The region between the edge of the continental shelf and the floor of the ocean.

bathymetry Measurement and mapping of the floor of the ocean by means of multiple depths measurements.

bathyscaph A submarine designed for exploration of the depths of the ocean.

bathythermograph A scientific instrument that can be towed behind a ship and that is used to measure and record temperatures at ocean depths down to about 980 feet (300 m).

bauxite The main ore from which aluminum is extracted.

bay A wide inlet in a sea or lake, smaller than a gulf.

bay beach A coastal beach in a bay between headlands. The headlands help to protect the beach's sand or shingle from being washed away by waves and currents.

bayou A very slow-moving, partly closed waterway that results when a river cuts a new channel for itself, leaving the old channel to silt up.

beach The sloping strip of coastal land between high and low water marks.

Beaufort scale A scale using the visible effects of a wind to measure its force, from 0 (calm) to 12 (hurricane). British naval officer Sir Francis Beaufort devised the first version in 1806.

bed (1) A layer of sedimentary rock more than 1/2 inch (1 cm) thick. (2) The floor of a sea, lake, or stream.

bedding The arrangement of sedimentary rocks into distinct layers. Different types of bedding help to indicate whether sediments were laid down in a stream, lake, sea, or desert.

bedding plane The surface between two beds of sedimentary rock.

bed load The heavy material debris that is carried along the bed of a river by the flow of the current. Bed load moves by sliding, rolling, or by small jumps (saltation).

bedrock The mass of solid rock that lies beneath the surface layer of unconsolidated and weathered rock material (the regolith).

Benioff zone A plane that dips downward at an angle from near the surface of the Earth's crust to a maximum depth of about 400 miles (700 km) and is thought to indicate active subduction. The Benioff zone forms the location of the start point of earthquakes.

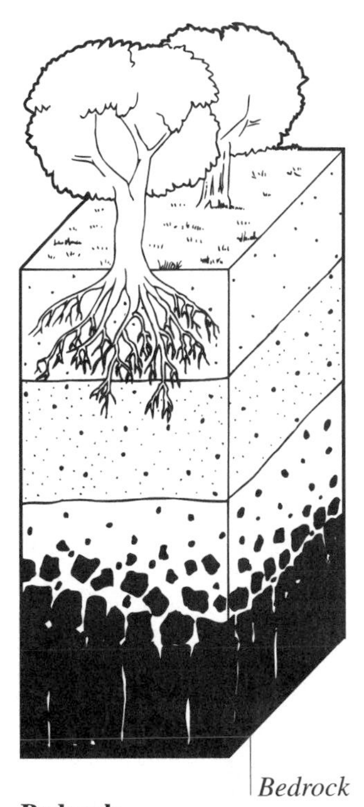

Bedrock

benthic Of the floor of the ocean or of any extensive body of water.

benthos Organisms that live on the floor of the sea. Many depend for food on dead plankton drifting down from the sea surface.

bentonite A clay, formed from volcanic ash, that has the property of swelling greatly in volume as a result of its ability to absorb water.

bergschrund A deep crevasse in a glacier filling a cirque (corrie). It forms where moving ice separates from ice fixed to the headwall.

berm A sand or shingle shelf high up on a beach, formed from material thrown there by waves during storms.

Beryl A hard mineral, an aluminum silicate of beryllium, used as an ore of beryllium and as a gemstone.

bifurcation Branching or separation of a stream.

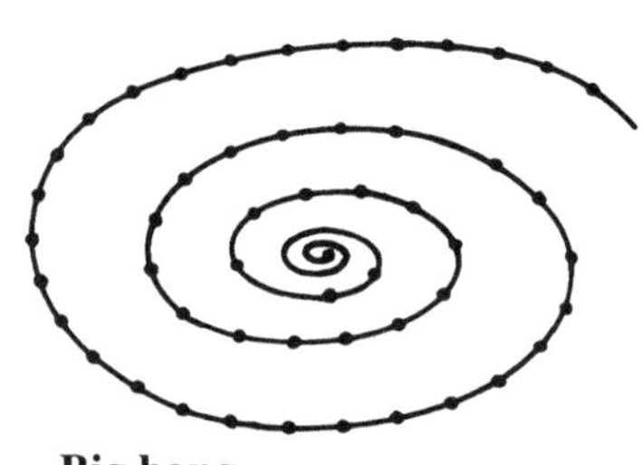

Big bang

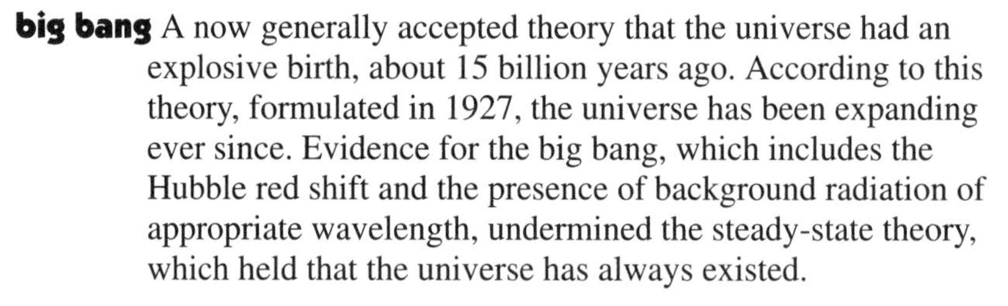

big bang A now generally accepted theory that the universe had an explosive birth, about 15 billion years ago. According to this theory, formulated in 1927, the universe has been expanding ever since. Evidence for the big bang, which includes the Hubble red shift and the presence of background radiation of appropriate wavelength, undermined the steady-state theory, which held that the universe has always existed.

binary star system Two stars held together by gravitational attraction and orbiting their common center of mass.

bioclast Any broken fragment (clast) that originated in an animal skeleton.

bioconcentration The concentration of pollutants, especially pesticides, in the living tissue of organisms at the top of a food chain.

biodegradability The ability of enzymes provided by bacteria to decompose and break down substances to simpler substances or elements. All organic matter is ultimately biodegradable, but many artificial materials are not and can thus cause environmental damage. Some plastics are biodegradable.

biogenic Arising from the remains or products of plants or animals.

biogenic deposit The formation of rocks, fossil traces, or other structures as a result of the earlier presence of living organisms.

biogenic sediment Any sediment formed from the skeletal or shell remains of formerly living animals or plants. A substantial proportion of the content of sedimentary rocks results from biogenic sediment.

biogeography The scientific study of where species live and why they live where they do. Such studies range from local to global in scale.

bioleaching Extracting a metal from low-grade ore by subjecting the ore to water and certain bacteria. *See also* leaching.

biomass The amount of organic matter in a given area or volume. Biomass fuels include firewood, dried dung, and biogas.

biosphere The Earth's thin outer zone where life flourishes. The biosphere chiefly comprises the lower atmosphere, land surface, soil, and fresh- and salt water.

biostratigraphic unit A rock layer (stratum) or group of strata that are identified and related to each other by fossil content.

biotite A green, brown, or black mineral; an aluminum silicate of potassium, magnesium, and iron of the mica group, found as a common constituent of metamorphic rocks and intrusive igneous rocks such as granite, as well as in sedimentary rocks.

bittern One of numerous soluble chemicals that are precipitated from seawater only in the final stages of evaporation.

bituminous coal A soft, black form of coal.

black hole In space, a vast mass compressed into a small area. Black holes exert a gravitational pull so strong that not even light can escape.

blizzard A fierce, bitterly cold wind with visibility reduced by wind-driven snow. Antarctic blizzards are among the severest of all.

block mountain A mountain mass squeezed between two parallel faults or left standing after lands flanking the faults have subsided.

bog Spongy, waterlogged ground comprising a mass of living and decayed mosses and other plants.

bolson A large, undrained basin in an arid or desert region surrounded by mountains.

bomb A clot of lava ejected by a volcano that takes a rounded form during its passage though the air and forms a cracked surface on cooling.

boracite A mineral consisting mainly of magnesium borate.

borate mineral A mineral in which the element boron is combined with oxygen and other elements such as silicon, arsenic, and phosphorus.

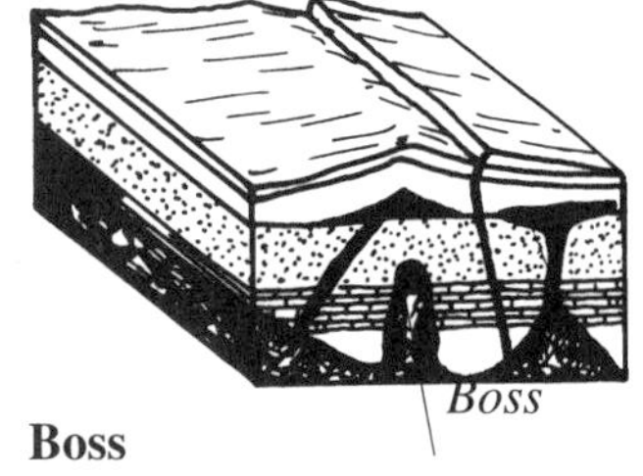

Boss

borax The mineral hydrated sodium borate.

bornite The mineral copper iron sulfide, which is valued as a copper ore.

bort The black diamond variety known as carbonado.

boss A rounded mass of intrusive igneous rock less than 15 miles (25 km) across.

botryoidal Of any mineral aggregate that resembles a bunch of grapes.

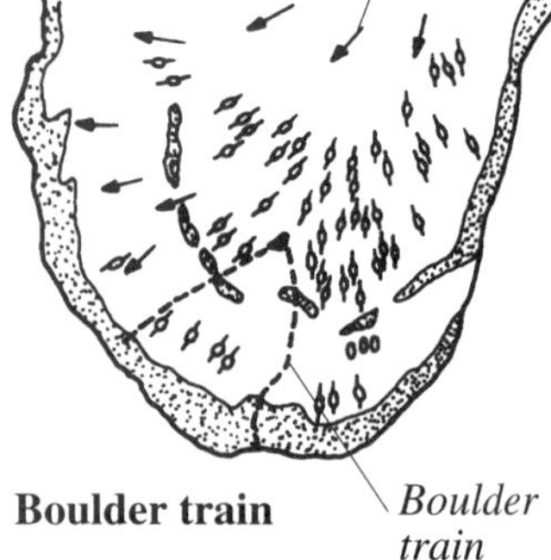

Boulder train

boudinage A structure in metamorphic and sedimentary rocks resembling a string of sausages.

boulder (1) A rounded lump of stone larger than a cobble. (2) A piece of gravel more than 10 inches (25.6 cm) across. *See also* cobble.

boulder clay *See* till.

boulder train A fan-shaped area of rocks spread from an outcrop by the movement of a glacier. The farther from the source, the smaller are the rock fragments.

braided stream A shallow stream separated by bars into a number of streams that rejoin in places downstream. Braided streams are common near ice sheets and glaciers.

breaker A wave that breaks on a shore after rising and toppling forward on reaching the shallows. A sharp rise in seabed promotes plunging breakers that crash forcefully down on the shore.

breccia A coarse sedimentary or igneous rock consisting of angular broken fragments (clasts) held together by a natural cement. Sources of breccias include scree slopes (taluses), faulting, volcanic eruptions, and meteorite impact.

breccia flow The movement of breccia in a lava flow. The fragments may be torn from the volcano chimney or collected from the land over which the lava moves.

breccia pipe An elongated, almost vertical, pipe-like structure of irregular broken fragments of rock.

brecciola Beds of small limestone breccia. The term is the Italian diminutive of breccia.

breeze Any wind of force 2–6 on the Beaufort scale (4–31 mph). The daily land and sea breezes in coastal areas result from the unequal rates at which sunshine heats land and sea.

brine Salty groundwater, often found in and around rocks with a high salt content.

brown coal Coal formed from peat. It represents an early stage in coal development between peat and bituminous coal.

brucite The mineral magnesium hydroxide occurring usually in plate-like crystals.

Bryozoa A phylum of simple marine animals that form lace-like branching or spiral colonies on the surface of shells, rocks, and other objects.

bund (1) An artificial ridge between fields.
(2) An embankment, dam, or quay.

butte A small, steep-sided hill with a flat top, formed when part of a mesa erodes away, leaving the butte, which is protected from

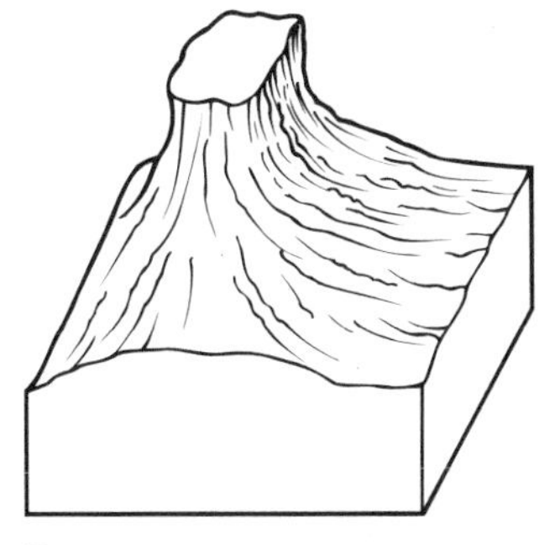

Butte

erosion by a resistant hard top rock. A butte is smaller than a mesa and a mesa is smaller than a plateau. All three have a similar structure. *See* mesa.

calaverite A mineral of the compound gold telluride. Gold most commonly occurs as the native metal.

calcareous Composed of, or relating to, calcium carbonate.

calcite A form of calcium carbonate that is a major ingredient in chalk, limestone, and marble.

calcium A silvery white metallic element. It occurs naturally only as part of a chemical compound.

calcium carbonate A chemical compound consisting of carbon and oxygen, and occurring as aragonite and calcite.

caldera A huge bowl-shaped volcanic crater, typically several miles across and with a steep inner slope. Calderas result from the explosion or subsidence of a volcano's original cone.

caliche A dull calcite deposit, derived from groundwater, that occurs in some areas of low rainfall.

Cambrian period The first period (540–505 million years ago) of the Paleozoic era. Multicellular organisms originated and became plentiful in Cambrian times.

canyon A deep, steep-sided, long valley cut by running water in a region that is otherwise dry. Rivers crossing dry regions form canyons by cutting down through the rocks. Rivers crossing once exposed continental shelves cut what are now submarine canyons scarring continental slopes. Ocean canyons on continental slopes are also believed to be cut out by local currents.

cape A pointed mass of land jutting into the sea.

cap rock A large boulder balanced, often precariously, on top of a slender pillar of rock, resulting from the erosion of the softer rock under the boulder.

carbon A nonmetallic element, noted for the enormous number of compounds it can form because of its valency patterns and the many ways carbon atoms can link in chains or rings. Carbon is essential to life and is the basis of the chemistry of almost all living organisms. Organic chemistry is the chemistry of carbon. Pure carbon occurs as the minerals diamond and graphite and as the powdered carbon black. In geology, carbon is important for the carbonates it can form with metals, as in the calcium carbonate of limestones and dolomites.

carbonaceous Of rocks that contain carbon.

carbonaceous chondrite A dull, black, stony meteorite, rich in carbon but low in metals.

carbonado *See* bort.

carbonate minerals Minerals that are a major feature of sedimentary rocks and are found mainly in limestones and dolomites, the most abundant being calcium carbonate or calcite. Aragonite has the same formula as calcium carbonate but is less stable. Dolomite is a carbonate of calcium and magnesium.

carbonates Compounds containing chemically combined carbon and oxygen atoms. Carbonates are the main ingredients in limestone, dolomite, and aragonite.

carbon cycle The circulation of carbon through the biosphere. Plants use photosynthesis to convert atmospheric carbon dioxide to carbohydrates, thereby making food that is eaten by animals. Breathing, burning, and decay return carbon dioxide to air.

carbon dioxide A gas made up of chemically combined carbon and oxygen atoms. It is produced in large quantities when organic matter is burned and is constantly present in the Earth's atmosphere. Rising levels of carbon dioxide increase the blanketing effect of the atmosphere and can lead to a rise in the mean temperature at the Earth's surface. Carbon dioxide is one of the "greenhouse gases." *See* greenhouse effect.

Carboniferous period The penultimate period (360–286 million years ago) of the Paleozoic era when enormous forests provided the raw material for the formation of great deposits of coal, hence the name. The Carboniferous is divided into the Mississippian and Pennsylvanian periods in North America.

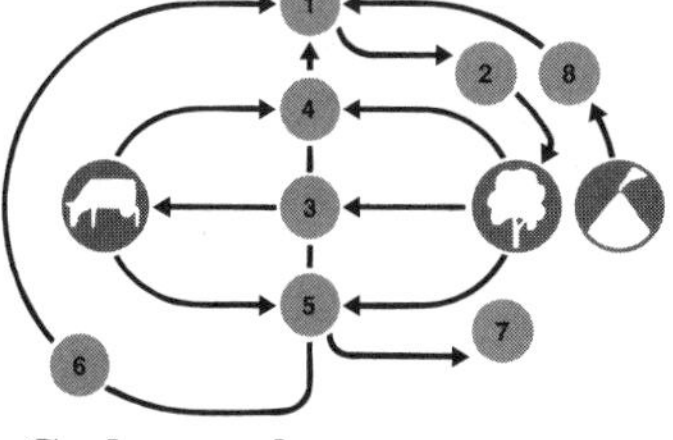

Carbon cycle
1 Carbon dioxide (CO_2) is in the air
2 Absorbed by plants
3 Animals eat plants
4 Breathed out by plants and animals
5 Bacterial decomposition of dead animals and plants
6 Carbon compounds converted into CO_2 by bacteria
7 CO_2 produced by burning fossil fuels
8 CO_2 produced by volcanic eruptions

carbonaceous – Carboniferous period
GLOSSARY

Carlsberg Ridge The gradually extending oceanic ridge that separates the African lithospheric plate from the Indo-Australian plate.

carnotite A strongly radioactive, bright yellow mineral consisting of the hydrous vanadate of uranium and potassium. It is a commercially important uranium ore.

cassiterite The reddish brown to black mineral tin oxide and the only important ore of tin.

casts Surviving reproductions of formerly living things produced when the impression of an organism forms a mold in clay or other material before it hardens and in which a positive mineral reproduction can form.

cataclasite A metamorphic rock formed when solid rock is physically changed by powerful crushing or grinding forces.

cataclastic metamorphism The process by which cataclasite is formed.

catastrophism A theory produced to explain the differences in fossils in different layers of sedimentary rock and to avoid conflict with biblical statements. The theory proposes that living things at each level were destroyed by a natural catastrophe and new organisms formed.

cave A hole in the Earth's crust produced by flowing water or lava.

Cave

celestial sphere An imaginary heavenly sphere surrounding the Earth and with the stars fixed to its surface.

celestite A mineral of strontium sulfate, commonly found as well-formed crystals, and the main source of the element strontium.

cement The material, such as calcite, that binds sedimentary grains together to form a rock. Cementing materials are deposited from mineral-rich water that percolates through pore spaces in rock.

cementation (of rock) The hardening of sediments into sedimentary rock by precipitated minerals filling tiny gaps between particles and cementing them together. Natural cements include silica, calcite, and iron oxides.

Cenozoic era The geological era following the Mesozoic era. It began about 65 million years ago and continues today. The Cenozoic is sometimes known as the Age of Mammals.

centrosphere The central zone of the Earth, under the crust and consisting of the core and the surrounding mantle.

Cephalopoda A class in the phylum Mollusca that contains the squids, cuttlefish, octopuses, the extinct ammonites, and the members of the genus *Nautilus*, especially the pearly nautilus.

cerargyrite The soft mineral silver chloride that contains more than 75% silver.

cerussite The mineral lead carbonate. An ore mineral of lead.

Cetacea The order of marine mammals that contains whales. The largest known animal is the blue whale.

CFCs *See* chlorofluorocarbons.

chabazite A zeolite mineral of hydrated aluminosilicate in which water molecules are held in cavities in the crystal lattice.

chalcanthite The mineral hydrated copper sulfate.

chalcocite The mineral copper sulfide, which is an important ore of copper.

chalcopyrite The mineral copper iron sulfide and an important ore of copper.

chalk An extensively deposited, white, fine-grained form of limestone rock consisting of the fossilized remains of marine plankton (coccoliths). Chalk may be almost 100% calcium carbonate.

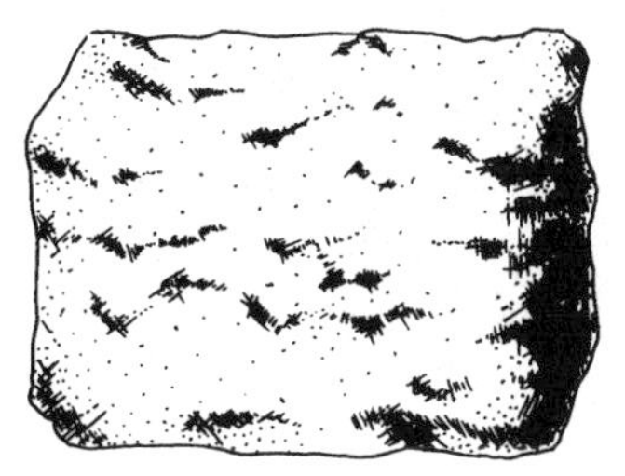
Chalk

channel (1) A navigable waterway.
(2) A strip of water broader than a strait and linking two stretches of sea.
(3) The course of a stream confined between banks.
(4) The bed and sides of any watercourse of whatever size.

chart (1) A visual presentation of data, such as a pie chart in which each slice represents one continent's population as a

percentage of world population, or a bar chart in which the length of each bar represents the level of rainfall in a particular month.
(2) A map designed for a special purpose, such as a meteorological chart or a navigational chart.

chatoyancy A property of some minerals of showing, in reflected light, an appearance like that of a cat's eye. Chatoyancy is well displayed in chrysoberyl.

chemical dating A method of estimating the age of a substance that undergoes a characteristic chemical change at a known rate.

chemical evolution The natural processes involving lightning discharges that are believed to have led to the formation of organic molecules such as amino acids, their elaboration into proteins and other compounds such as RNA and, by means of the recently discovered enzymatic properties of RNA, the production of the first simple life-forms, from which all subsequent living organisms evolved by natural selection.

chemical weathering The breakdown of rock by chemical processes. For instance, rainwater with dissolved carbon dioxide dissolves limestone and removes the natural cements from some other rocks.

Unweathered granite mass

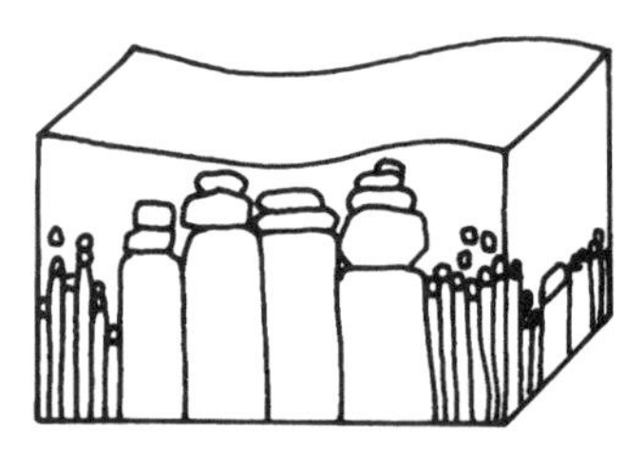

After substantial weathering

Exposed massive granite blocks (tors)

Chemical weathering

chert A sedimentary rock made of tiny quartz crystals.

Chiroptera The order of flying mammals, consisting of the bats.

chlorite One of a group of mainly greenish minerals of phylosilicate with a layered structure like that of the micas.

chlorofluorocarbons (**CFCs**) Compounds of chlorine and fluorine once much used as aerosol propellants and refrigerants and in foam packaging. In the Earth's atmosphere CFCs are now known to act as catalysts in a reaction that breaks down ozone, thus depleting the Earth's ozone layer and increasing the greenhouse effect.

chlorophylls The pigments that confer the green color on plants and absorb light, thereby securing the energy needed for the conversion of carbon dioxide and water to sugar and starch (photosynthesis) and the initiate of the food chain. Chlorophyll *a* is the primary pigment. Chlorophylls *b*, *c*, and *d* are accessory pigments that pass their energy to chlorophyll *a*, where photosynthesis is initiated.

Chlorophyta Green algae that contain chlorophylls and closely resemble plants.

chloroplasts Lens-shaped, chlorophyll-containing cell organs (organelles) present in large numbers in the cells of plants undertaking photosynthesis.

chondrite A meteorite containing small rounded bodies of olivine and enstatite embedded in olivine and other minerals. Chondrites are the commonest kind of meteorite.

chromite A mineral of iron oxide and chromium that is the only usable ore of chromium.

chromosphere The lowest level of the Sun's atmosphere, above the photosphere and below the corona.

chrysoberyl The mineral beryllium aluminum oxide.

chrysotile The fibrous form of the mineral hydrous magnesium silicate (serpentine), which is the main source of asbestos.

cinder cone A massive conical structure with a cup-shaped upper opening formed when molten rock, ejected vertically by a volcano vent, cools and falls back down again as black gravel.

cinnabar The mineral mercury sulfide and the only commercially important ore of mercury.

cirque A mountain rock basin scooped out by freezing, thawing, and slipping ice formed from compacted snow. Many cirques hold a small, round lake. *See also* tarn.

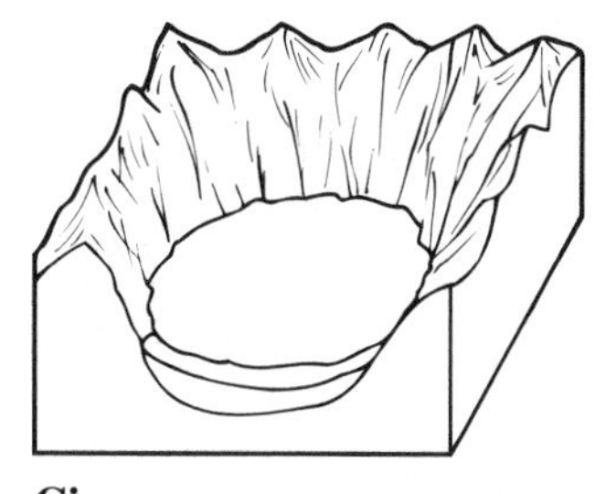
Cirque

cirrocumulus A rippled, high-level cloud made of ice crystals. Blue sky can be seen between the ripples.

cirrostratus A veil of high cloud covering the sky like a thin veil. Its ice crystals may produce a halo around the Sun or Moon when these are seen through this cloud.

cirrus Wispy, high-level clouds made of ice crystals. They are often drawn out into "mares' tails" by powerful winds.

citric acid cycle Another name for the Krebs cycle, by which energy is transferred from the oxidation of food materials to ATP so that it can be made available for all the complex metabolic processes occurring in living cells.

cladogenesis *See* adaptive radiation.

clast Any weathered fragment of rock.

clastic Consisting of broken rock fragments or organic residues that have been transported from their point of origin.

clastic rock Rock formed from the broken fragments of older ones. Clastic rocks include conglomerates and sandstones.

clay A fine-grained sedimentary rock made of tiny, flaky crystals. Clays tend to be plastic when wet and hard or powdery when dry. They are porous but impervious.

clay mineral Any of a group of chemically related aluminum silicates of very small crystal size that form a fine-grained mass that remains plastic when wet but becomes hard when allowed to dry or when heated (fired). The most important clay minerals belong to the kaolinite, chlorite, hydrous mica, smectite, talc, and vermiculite groups.

Climate

Polar

Temperate

Tropical

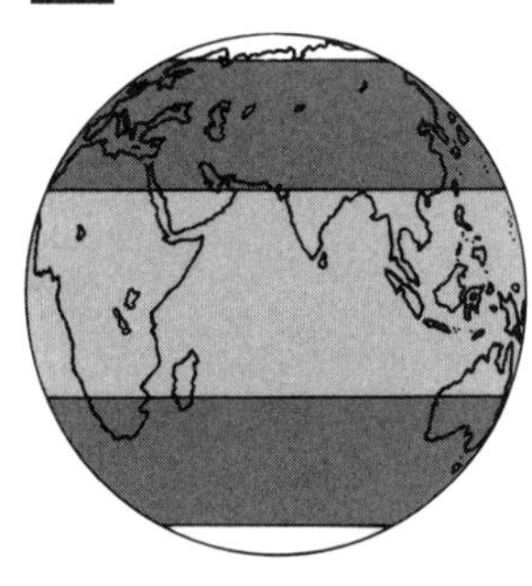

cleavage (1) The tendency of some sedimentary and metamorphic rocks to split along parallel lines of weakness.
(2) The tendency of crystals to split along planes of weakness determined by the crystals' molecular structure.

cliff A very steep rock face, such as sea cliffs and gorges.

climate The average weather of a region, locality, or larger area as measured for all seasons over a number of years.

climatology The scientific study of climate. Climatology covers much more than simply the description of climate and climatic change; it also deals with climate's origins and practical consequences. Thus, it overlaps with a range of other sciences, such as oceanography, geology, geophysics, biology, agriculture, engineering, and economics.

clints Limestone ridges separated by gullies. Clints form where weathering erodes a horizontal limestone surface. *See also* grikes, karst.

cloud A mass of water droplets or ice crystals in air, formed by water vapor droplets condensing, usually high above ground, as a result of lowered atmospheric temperature. There are three basic cloud forms: cumulus, cirrus, and status. Cumulus are the dense, billowy clouds rising in great domes or towers from a level low base. Cirrus are high white clouds with a silky or fibrous appearance. Stratus clouds are extensive, low, flat layers or patches without detail. Combinations of these three types produce a range of other descriptions, such as cirrostratus, cirrocumulus, cumulonimbus, and stratocumulus. Other terms relate to altitude and effect.

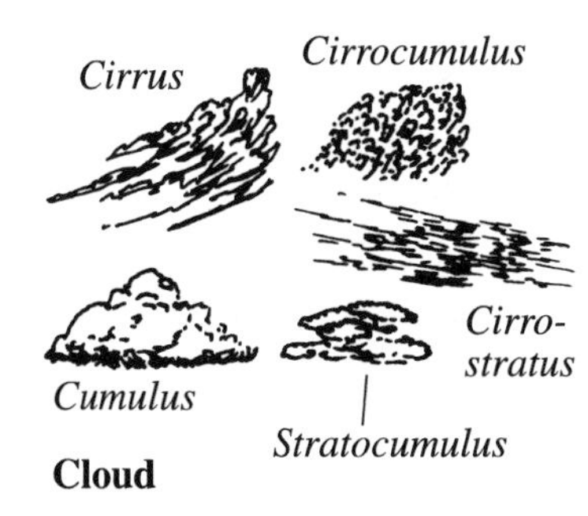

Cloud

coal A carbon-rich rock that burns easily. It consists of layered plant remains compacted by pressure over millions of years. Coal varies in hardness and calorific value, depending on the conditions of its formation; it ranges from anthracite, through bituminous coal, sub-bituminous coal, and lignite (brown coal). *See also* anthracite, bituminous coal, brown coal, peat.

coast *See* coastline.

coastal plain A low-lying, flat stretch of land bordering a sea. Coastal plains include sediments deposited by rivers and seafloors exposed by a fall in sea level or a rise in land level.

coastline or **coast** The edge of the sea, marked by sea cliffs or a line of debris dumped by storm waves high up on a beach.

cobaltite The mineral cobalt sulfarsenide, usually containing some nickel.

cobble A rounded stone between a pebble and boulder in size.

coccoliths Microscopic chalky plates or disks that form protective coverings for single-celled plankton organisms and that form a large part of the ocean ooze. Coccoliths are important components in sedimentary chalky rocks.

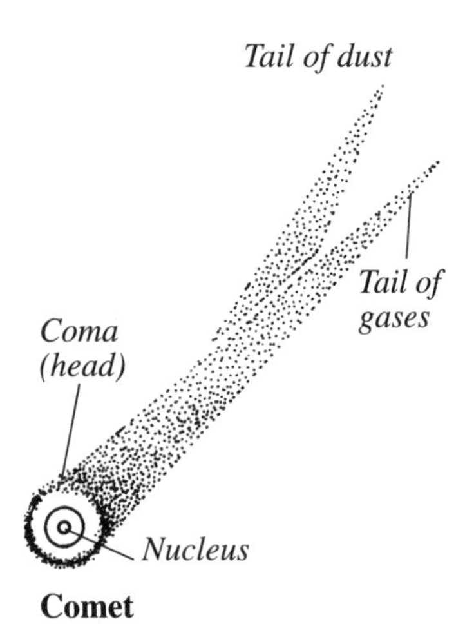

Comet

col A natural pass in a mountain range.

cold front A boundary between cold and warm air masses. An advancing wedge of cold, dense air forces the warm air to rise above it. Cold fronts bring showers and a drop in temperature.

comet A dust, gas, and ice body that orbits the Sun. On nearing the Sun it develops a long, bright tail.

commensalism The association of two different species of plant or animal without interdependence. Commensal relationships are not parasitic and cause no harm to either party. In commensal relationships, the advantage is invariably with the guest, which is furnished with a suitable habitat by the host, providing such benefits as support, shelter, transport, and food.

compaction The physical process brought about by the increasing pressure at the base of an ever-thickening sediment.

competition The ecological process in which a population of one species consumes any resource at the expense of a population of another, thereby causing harm to the latter. Competition may be concerned with food, water supply, oxygen, living space, or availability of mates. Two populations of different species in the same area, both being dependent on the same set of limited resources, are said to occupy the same niche. When this happens, one will tend to predominate and to exclude the other because the characteristics of one will always give it an advantage over the other in terms of obtaining the required resources.

compound A substance made of two or more elements chemically combined.

conchoidal fracture A break in a mineral that is smoothly concave or convex and that resembles the surface of a shell.

concretion A mineral forming a hard nodule within a sedimentary rock. The flint nodules in chalk are concretions.

condensation The formation of a liquid from a vapor. In air saturated with water vapor, droplets of water form around condensation nuclei such as particles of salt, dust, and the like.

cone A hill or mountain that tapers from a broad round base to a narrow circular apex. Most volcanoes are cone-shaped.

conglomerate Any sedimentary rock containing smooth, rounded rock fragments (clasts) greater than 2 mm in diameter embedded in the fine-grained matrix, as of sandstone or limestone. Such a rock is known as a rudaceous rock.

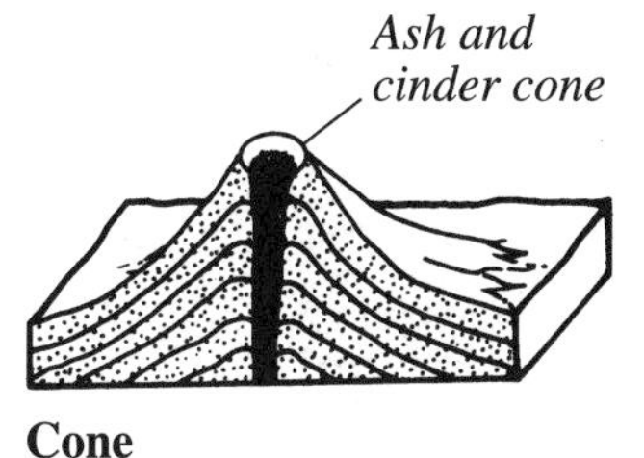

Cone

conic projection A map projection made as if lines of latitude and longitude project from a globe onto a paper cone that touches one or two lines of latitude and has its apex over one Pole. Conic projections can be designed to show distances accurately.

conservation Protection of the environment by careful management of natural resources.

constellation An apparent grouping of stars.

contact metamorphism or **thermal metamorphism** Metamorphism in igneous intrusions resulting from raised temperature alone.

continent One of the Earth's major landmasses. Continents occupy about 29% of the Earth's surface.

continental crust The material forming the continents. Continental crust is thicker and less dense than oceanic crust, and it contains more complex rocks.

continental drift The hypothesis that continents have moved relative to each other over the Earth's surface. This was based, initially, on the observation that the west coast of Europe and Africa would fit neatly into the east coast of the Americas. Although at first derided, the hypothesis has gained considerable support from the growing evidence that plate tectonics is a reality.

continental rim The rim of the continental shelf surrounding a continent and forming its true but submerged structural edge.

continental rise The gentle gradient at the foot of a continental slope. It consists of sediment that has slid down the slope.

continental sea A shallow sea that submerges a low-lying part of a continent. At times, continental seas have occupied much of North America.

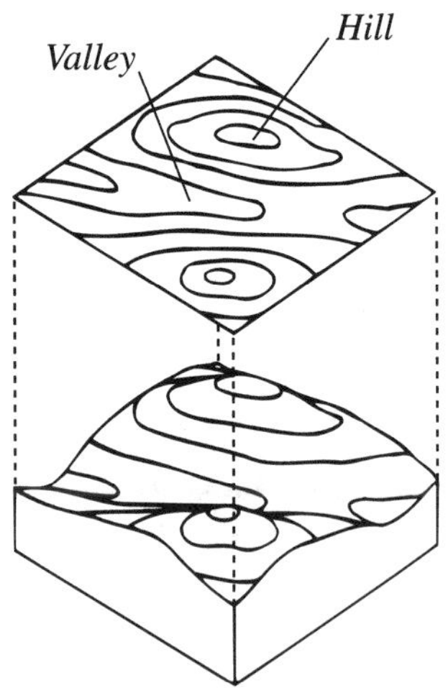

Contour maps

continental shelf The submerged edge of a continent, forming a shallow seabed offshore.

continental slope A relatively steep submerged slope descending from a continental rim to a continental rise or an abyssal plain.

contour map A map in which contour lines represent heights. Each contour line runs through places with the same altitude. The closer the contour lines, the steeper the slope between them.

contour plowing Plowing the soil along a slope instead of up and down it. Contour plowing reduces the risk of soil erosion on most types of land.

convection Heat transfer by movement of the medium heated, such as through the movement of air, water, or molten rock below the Earth's crust.

convection zone A region of upwelling matter below the surface of the Sun or another star.

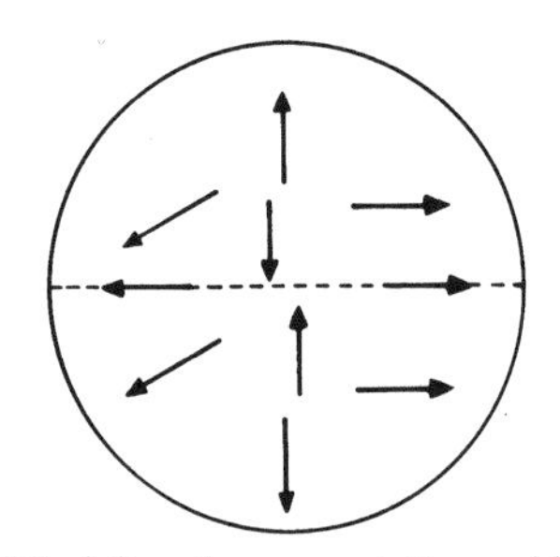
Wind direction on a static world

convectional rainfall Rainfall that occurs when heated moist air rises, expands, and cools until its moisture condenses to form clouds and rain.

coral reef Coral forming a shallow, submerged rock platform, with a steep outer slope and a gentler inner slope. Much of the world's limestone rocks originated as prehistoric coral reefs.

cordillera A system of parallel mountain ranges with plateaus and basins between them, such as the ranges and plateaus of western North and South America.

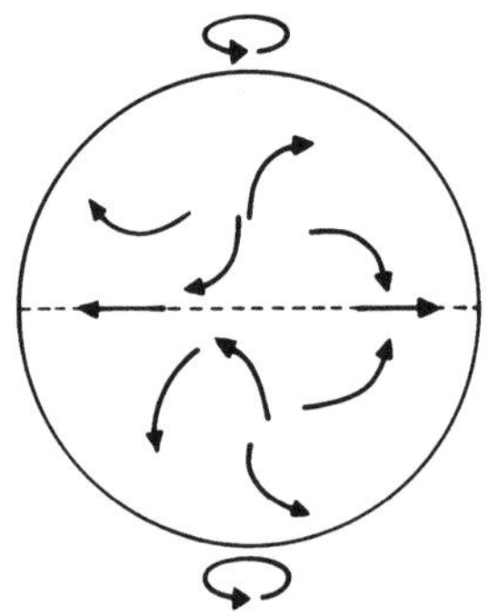
Deflections of wind caused by Earth's spin
Coriolis effect

core The dense matter forming the center of a star, planet, or Moon. The Earth's outer core, 1,400 miles (2,240 km) thick, is of molten iron and nickel, with some silicon. The inner core, 1,540 miles (2,440 km) across, may be iron and nickel at 6,700°F (3,700°C). Extreme pressure prevents it from becoming liquid.

Coriolis effect The tendency of the Earth's rotation to deflect winds and currents to the right in the Northern Hemisphere and to the left in the Southern Hemisphere.

corona The Sun's gaseous outer layer, visible as a halo during a total solar eclipse and emitting a so-called solar wind of energetic particles.

corrasion The abrasive erosion of a solid rock surface by particles dragged over it by river water, glaciers, sea waves, or the wind.

corrie *See* cirque.

corrosion The chemical wearing away of solid rock by natural acids, etc. *See also* weathering.

cosmology The study of the universe, especially its origin and evolution.

country rock Any preexisting rock into which igneous intrusion occurs or which has been surrounding a mass of igneous rock.

covalent bonding A chemical bonding between atoms sharing electrons.

crag and tail A type of hill found in glaciated areas. One side is a short, steep slope of hard rock (the crag); the other side is a long, gentle slope of soft rock (the tail). The crag resisted glacial erosion and protected the tail.

crater (1) A circular depression, usually with a raised rim, produced by meteorite impact on a planet or moon.
(2) A funnel-shaped hole in a volcanic cone, produced by volcanic eruption.

crater lake A lake occupying the crater of an extinct or dormant volcano.

craton An ancient part of a continent that has not been deformed by mountain-building activity.

creep The gradual downslope movement of soil or rock fragments, caused by gravity. Creep is one of the forms of mass-wasting.

Cretaceous period The last period (144–65 million years ago) of the Mesozoic era. Continents drifted apart and dinosaurs diversified greatly before becoming extinct.

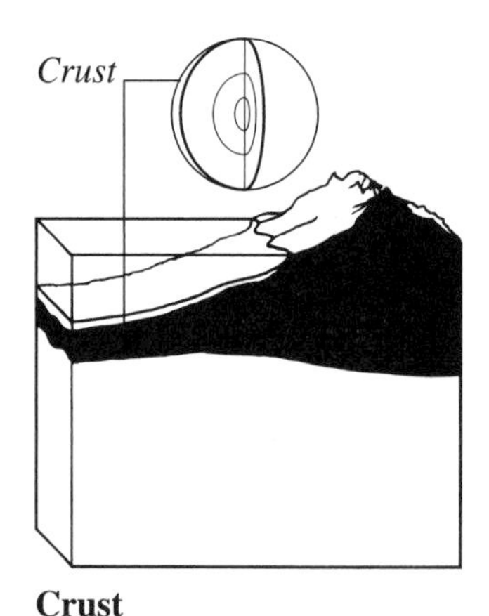

Crust

crevasse A deep crack in a glacier. Crevasses form where the valley floor suddenly steepens or widens.

crop rotation Growing different crops in alternate years. Crop rotation helps preserve soil fertility.

cropland Arable farmland.

crust The hard outer layer of a solid-surfaced planet or moon. The Earth's continental crust averages 20 miles (33 km) thick. Oceanic crust, forming the ocean floor, is less than 6.2 miles (10 km) thick. The Earth's crust consists mainly of crystalline rock and extends from the surface to the Mohorovičić discontinuity.

crystal (1) A substance that has solidified.
(2) A solid of definite chemical composition and with a characteristically ordered geometrical arrangement of atoms and faces that form naturally with external symmetry.

cuesta A ridge with a steep scarp on one side and a gentle dip slope on the other. *See also* escarpment, dip slope.

cumulonimbus A type of towering cloud with a low base but a high top, often spread out in the shape of an anvil. Cumulonimbus clouds often generate thunderstorms.

cumulus A type of cloud with a flat base and rounded or cauliflower-shaped top. Cumulus clouds form in warm, sunny weather but sometimes develop into stormy cumulonimbus clouds.

cycle A repeated sequence of events. *See* carbon cycle; nitrogen cycle; oxygen cycle; water cycle.

cyclone An atmospheric circulation system in which the wind rotation direction is the same as that of the Earth's rotation. Thus, a cyclone rotates clockwise in the Southern Hemisphere and counterclockwise in the Northern Hemisphere. The Coriolis force, due to the Earth's rotation, is directed to the right of the flow in the Northern Hemisphere. The pressure gradient force, which is directed toward low pressure, acts in the opposite direction. These two forces together bring about an atmospheric pressure minimum (low or depression) at the center of the cyclone. The result is usually windy, rainy, unsettled weather. Violent tropical cyclones, on the other hand, get their energy from the release of the latent heat of condensation in precipitating cumulus clouds; over the oceans it can result in hurricanes and typhoons.

cyclonic rainfall Rainfall brought by the fronts of a passing atmospheric low or depression. After the drizzle associated with its warm front comes the cold front's showery conditions.

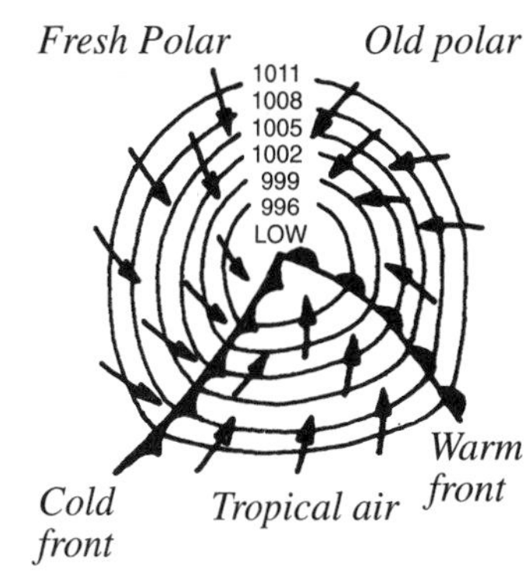

Cyclonic rainfall

cylindrical projection A map projection made as if the Earth's surface features were projected from a globe onto a rectangular sheet of paper curved around it to form a cylinder. Cylindrical projections greatly exaggerate the sizes of polar regions. *See also* Mercator projection.

dam A wall or mound blocking a river and stemming or diverting its flow. Artificial dams provide water for irrigation, hydroelectricity, navigation, and flood control. Natural sediment dams create oxbow lakes.

dating Calculating the ages of rocks or fossils. Various methods are used. *See* dendrochronology; paleomagnetic dating; radiocarbon dating; radiometric dating.

Dam

day (1) The length of time for which the Sun shines on one part of the spinning Earth.
(2) The 24 hours beginning at midnight.

Decapoda The order of crustaceans that includes shrimps, prawns, crayfish, lobsters, and crabs. They have five pairs of walking legs, of which the first pair form powerful pincers. The term is also used for the order of cephalopods that includes the squids and cuttlefishes.

decay The spontaneous change of a radioactive element into another element with the emission of particles or photons. In the process, the atomic number may rise or fall. The half-life of the element is the time taken for half the element to decay.

deciduous Of plants that shed their leaves each growing season. By limiting transpiration (water loss), this helps them to retain water. Most woody plants in temperate climates are deciduous.

declination A star's angular distance in degrees north or south of the celestial equator: the astronomical equivalent of terrestrial latitude.

deflation The windblown transportation of dust and sand in dry lands. Dust and sand scour some desert surfaces, but dust blown from deserts and the edges of ice sheets can settle as fertile deposits called loess. *See also* deflation hollow, loess.

deflation hollow A depression scoured in a desert surface by windblown sand and dust.

deforestation Clearing land of a forest by burning or chopping down trees. Most deforestation occurs where people replace forests with farmland.

deformation Faulting, folding, tilting, or other changes in rocks affected by movements in the Earth's crust.

delta An area of low, flat, often fan-shaped land at a river mouth. A delta is built out into a lake or sea by sediments dumped by a river that divides many times, crossing the delta as streams called distributaries.

Dendritic drainage pattern

dendritic drainage pattern A river system where the main river and its tributaries resemble the trunk and branches of a tree.

dendrochronology A method of dating using changes in the annual growth rings of trees in the same area caused by local climate changes. Wide rings occur in rainy seasons when growth is rapid; narrow rings occur in dry seasons.

denitrification The release of gaseous nitrogen from soil nitrates into the atmosphere, under the influence of bacteria. This is the opposite process from nitrogen fixation.

density A substance's mass per unit volume.

density current Deep ocean currents caused by differences in water density resulting from differences in salinity, temperature, or suspended matter content. More dense water will move downward under gravity to form deep currents.

denudation The wearing away of the land surface by processes such as weathering and erosion.

deposition The laying down of dust, mud, sand, gravel, or other materials removed by weathering and transportation. Most deposition results from the action of rivers.

depression (1) In meteorology, an area of low atmospheric pressure. *See* cyclone.
(2) In geomorphology, a dip in the Earth's surface, as formed by a deflation hollow or a syncline.

Dermaptera The order of insects that comprises the earwigs.

desalination Removal of salt from seawater so that it can be used for agricultural and drinking purposes. This may be done by filtration or by an ion-exchange process.

desert A region too dry to support many plants. Deserts have been defined as places with less than 10 inches (25 cm) of rainfall a year. Polar regions are technically deserts.

desertification The turning of fertile land into unproductive desert through land mismanagement.

detritus Loose mineral matter worn off rocks.

devitrification The change from a vitreous to a crystalline state.

Devonian period The fourth period (408–360 million years ago) of the Paleozoic era, when trees, insects, and amphibians appeared.

dew point The temperature at which water vapor saturates air. Below it, the condensing vapor forms droplets of water.

diabase or **dolerite** A dark, hard, medium-grained intrusive igneous rock found in dikes, sheets, sills, etc.

diagenesis The physical and chemical changes that occur in a sediment at normal temperatures and pressures to turn it into a rock. Diagenesis includes compaction, crystallization, cementation, dissolution, and replacement. With increasing temperatures and pressures, diagenesis merges into metamorphism.

diamond An allotropic form of pure carbon, crystallized as cubes or octahedra. Diamond is the hardest known mineral. It rates 10 on the Mohs' scale of hardness.

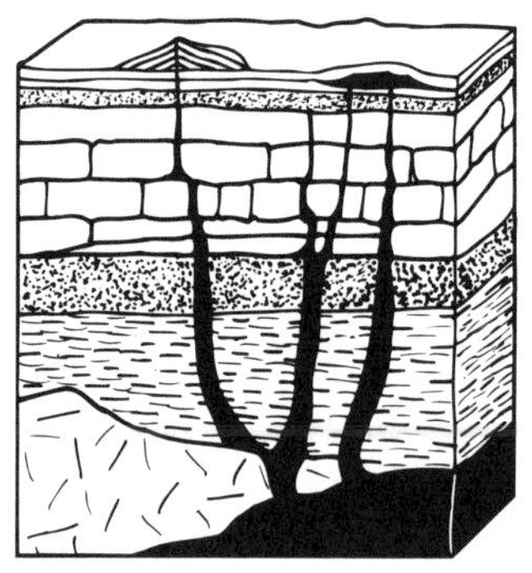
Dike

Dictyoptera An order of insects comprising the cockroaches and the mantids.

dike A vertical sheet of igneous rock intruded through older rocks so that it cuts across their bedding planes.

dinosaur One of many extinct reptiles of the orders Saurischia and Ornithischia that flourished during the Mesozoic era. Many were of enormous size. The Saurischia mainly ran on two legs and were carnivorous. The Ornithischia and some of the Saurischia were quadruped herbivores. The causes of the extinction of the dinosaurs are still being argued. *See* Ornithischia, Saurischia.

diorites Medium- to coarse-grained intrusive igneous rocks consisting largely of plagioclase feldspar with up to 10% of quartz with pyroxenes or hornblende.

dip The angle between the horizontal plane and a tilted rock bed or fault.

dip slope A slope aligned with the underlying rock layers, especially the dip slope of a cuesta.

Diptera A large order of insects containing all the two-winged flies, including house flies, horse flies, gadflies, biting black flies, sandflies, mosquitoes, and midges.

disconformity An unconformity where erosion has left a gap in the sequence of rock strata but those above and below it are horizontal.

disphotic Of the ocean zone to which little or no light is able to penetrate.

dissection The process by which rivers cut valleys in a smooth land surface, for instance, turning a plateau into a maze of gorges and ridges.

distributary A stream diverging from a main river, as in a delta.

divide The boundary between two river basins. It is usually an upland region or ridge.

doldrums The low-pressure equatorial belt of calm air and light winds.

dolerite *See* diabase.

doline *See* sinkhole.

dolomite A calcite-like mineral made of calcium magnesium carbonate; also the limestone-like sedimentary rock of which it is the main component.

drainage basin *See* watershed.

drainage pattern The overall pattern of a river and its tributaries. Drainage patterns depend on such things as faults, folds, and the underlying rocks' resistance to erosion.

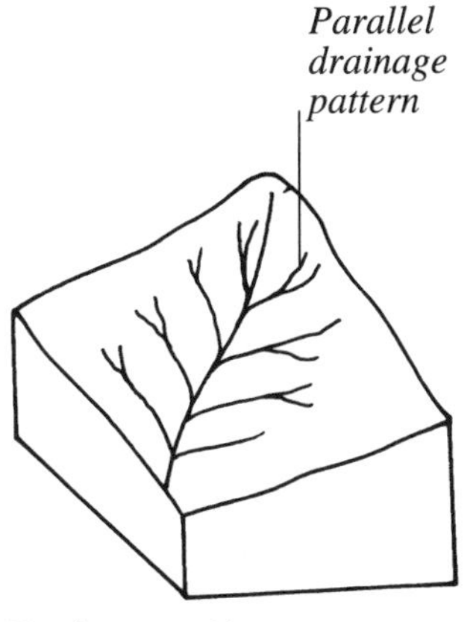

Drainage pattern

dredging Excavating gravel, sand, or mud from the bed of a sea, lake, or river.

drift Any rock material deposited by moving rivers, streams, or glaciers.

drift deposit Material dumped by a glacier or by streams flowing from a glacier or an ice sheet.

drift ice Ice floes or icebergs floating on the sea and borne by currents far from the place where they originated.

drift mining Driving a horizontal passage into a hillside to exploit coal or minerals, forming a seam or vein in the rocks.

drizzle Fine rain, often continuing for some time, falling from the stratus clouds associated with an atmospheric depression.

drought A long period of dry weather, defined in the United States as 21 or more days with not more than 30% of the average rainfall for that time of year. Drought causes serious hydrologic imbalance, with crop damage, water supply shortage, etc., in the affected area.

drowned valley A coastal valley drowned when the land sinks or the sea level rises. Drowned valleys typically form inlets in uplands. *See also* fjord, ria.

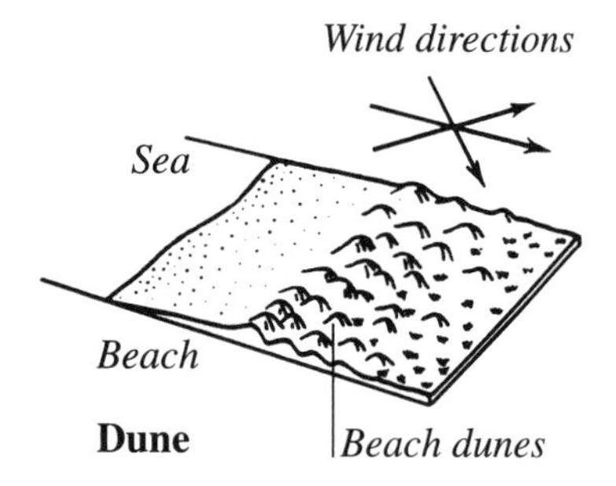

Dune

drumlin A smooth oval-shaped hill rounded at one end and tapered at the other so as to present a streamlined appearance. Drumlins may occur singly or in groups and are believed to be formed by an expanding ice sheet and streamlined by the movement of the ice.

dry valley A river valley without a river. Dry valleys are features of chalk and limestone hills where the surface was once partly frozen.

dune A sand ridge or mound formed where the wind heaps up sand in a desert or on a low sandy coast.

dyke A vertical or near-vertical sheet-like intrusion of igneous rock that has crystallized and settled just under the Earth's surface.

Earth Our solar system's fifth largest planet, the third in distance from the Sun. Its features include an oxygen-rich atmosphere, plentiful water, a hard crust, and an interior of hot rock and metal. The Earth is the only known planet with conditions that support life.

Earth sciences A wide range of disciplines concerned with all aspects of the Earth and its relation to the universe. The total content is a matter of dispute, but no critic would exclude geology, oceanography, meteorology, geophysics, and geochemistry. Most would also include cosmology, cosmogony, astronomy, and ecology, and some would include geography. With specialization, each of these is further subdivided, so the Earth sciences encompass many studies, such as paleontology, geomorphology, mineralogy, petrology, economic geology, volcanology, climatology, meteorology, oceanology, plate tectonics, paleoclimatology, paleoecology, paleogeography, stratigraphy, and the wide range of topics involved in ecology.

earthquake In popular terms, a sudden shaking of the Earth's crust. The term *earthquake* should strictly, however, be used to refer to the source of seismic waves, rather than the shaking, which is an effect of the earthquake. Earthquakes tend to occur in narrow, continuous belts of activity that surround large seismically quiet regions known as lithospheric plates. Plates are in continuous slow motion relative to each other, and as the plates move against each other, their edges do not slip smoothly but engage in a series of "slip-stick" sudden jerks. These jerks are earthquakes. Volcanic eruptions trigger some earthquakes.

eclipse The obscuring of the light from a celestial body when another body is interposed or casts a shadow upon it, cutting out light from the Sun. An eclipse of the Moon occurs when the Earth's shadow falls on it. An eclipse of the Sun occurs when the Moon is interposed.

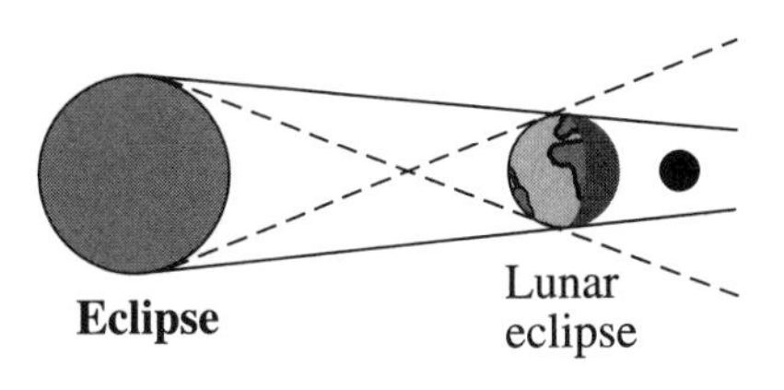

Eclipse

ecliptic The Sun's apparent path across the sky.

ecology The scientific study of how living organisms of all kinds interact with each, and the effects they have on each other and on their environments. Interactions may be brief or persistent and may continue for most of the lifetime of one or both of the organisms (inter-species associations). Some of these are damaging both to organism and environment; some are of benefit to one but destructive to the other; some are mutually beneficial. Associations may furnish, among other things, shelter, protection, living space, physical support, camouflage, a supply of essential nutrients, transportation, or a means of dissemination. *See also* mutualism; commensalism; competition; predation; parasitism.

economic geology The application of geological knowledge and skills to commercial purposes. Economic geology is important in prospecting for, and mining of, metals and minerals, and in the location and production of petroleum and natural gas. Many civil engineering projects also involve economic geology.

ecosystem A community of living things and their environment (pond, tree, forest, etc.).

electrical and electromagnetic exploration The study of the variations in electrical conductivity of, or electrical capacitance between, rocks for the purpose of prospecting for metallic ores. Measurements of both natural and induced electrical currents are made, and in the latter case both direct currents and low-frequency alternating currents are used. Radio-frequency studies from aircraft are also performed.

electron A subatomic particle with a negative electric charge.

element A chemical substance containing only one kind of atom. An element cannot be chemically broken down into a simpler substance.

ellipsoid A surface that can be cut in any direction to produce ellipses or circles.

El Niño ("The Child") A warm ocean current sometimes displacing a cold current off Peru with a disastrous impact on fisheries; it is part of a wider phenomenon with severe effects on tropical weather almost worldwide.

elongation index or **elongation ratio** The percentage by weight of mineral particles whose long dimension is more than 1.8 times the mean dimension.

ENSO Abbreviation for El Niño and Southern Oscillation (a periodic atmospheric pressure reversal between East and West Pacific preceding an El Niño event). *See* El Niño.

Eocene epoch The second subdivision (55–38 million years ago) of the Tertiary period.

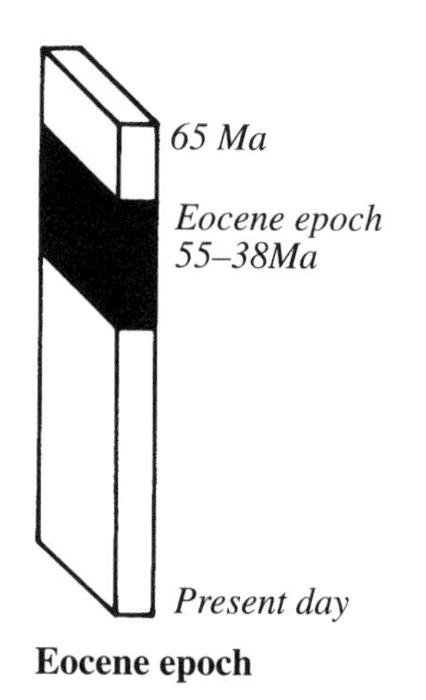

Eocene epoch

eolian Borne, eroded, or deposited by wind. The term describes wind-shaped features of landscapes, notably in deserts but also on some coasts and near the edges of ice sheets.

eolian deposits The settling of windblown sediments.

eolian process Erosion, transport, and deposition of material by the wind. Eolian processes occur most effectively in desert regions.

eolianite A general term for the products of eolian processes when they form a sediment.

eon The largest unit into which geological time is divided. Four eons have been identified.

ephemeral Describing a plant with a very short life cycle so that more than one generation can be created in one growing season. The term is also applied to animals, such as the mayfly, with a very short life.

epicenter The Earth's surface directly above an earthquake's focus or point of origin.

epoch A time unit forming a subdivision of a geological period.

equal area projection A map projection showing the relative areas of regions. Their shapes are liable to be distorted, however.

equator An imaginary line around the Earth midway between the North and South Poles.

equatorial plane An imaginary flat surface cutting through the Earth at the Equator.

equatorial zone The hot belt between latitudes 10° north and south of the Equator.

equinox ("Equal night") Either of two dates each year when day and night are of equal length everywhere on Earth. The vernal (spring) and autumnal (fall) equinoxes occur when the Earth's axis is at right angles to a line between the centers of the Sun and Earth.

era A time unit forming a major subdivision of geological time. An era contains at least two periods.

erathem The rocks formed during a geological era.

erosion The wearing down, loosening, and transportation of features of the surface of the land by wind, water, gravity, and moving land and ice. Erosion is a fundamental geologic process leading to the movement of material to lower levels and often its reconstitution into new sedimentary rocks. Erosion occurs both physically, without chemical action, and chemically, when rock materials have been dissolved or decomposed. Erosion is, to some extent, balanced by geologic processes such as volcanic eruption and plate tectonics, which raise the Earth's surface.

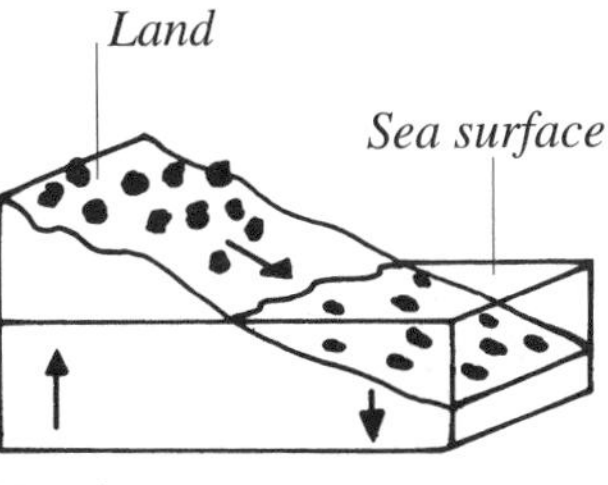

Erosion

erratic Large stones and boulders moved by a glacier and deposited at a site where such stones do not normally occur. Scottish granite found in Snowdonia in Wales is an erratic.

eruption An outpouring, especially of volcanic gas, lava, and ash from the Earth's surface.

escarpment A long continuous steep slope facing one direction between two more level areas.

esker A long, winding ridge of sand and gravel. It consists of sediments shed by a stream flowing under a stationary glacier that has since melted away.

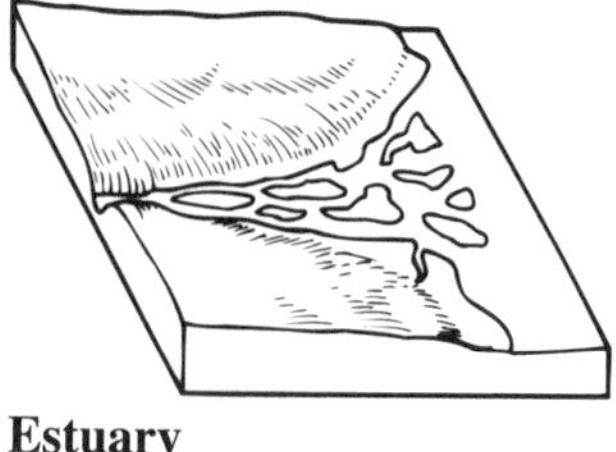

Estuary

estuary A broad, low, wide river mouth where it runs into the sea and where tidal salt water mixes with the fresh river water. Estuaries may occur when a low-lying valley drowns as the land sinks, or the sea level rises.

euhedral Describing crystals in igneous rock that show well-developed crystal faces.

euphotic Describing the ocean zone near enough to the surface to allow sufficient penetration of light to support photosynthesis.

eustatic Describing a world-wide change in sea level.

eutrophication Over-enrichment of water by nutrients (such as chemical fertilizer), causing plant overgrowth and decay, resulting in deoxygenation of the water and death of its organisms.

evaporation The change of a liquid into a gas. Water evaporating from oceans, rivers, and the land produces the water vapor in air.

evaporite Rock made of crystals precipitated by the evaporation of salt-saturated water.

evolution Processes of change, including those by which atoms form stars, planets, and living organisms that give rise to new species. *See also* chemical evolution.

exfoliation The splitting off from a rock of concentric thin sheets or shells from the outer layers. It occurs where daily heating and cooling cause rock surfaces to expand and contract. Exfoliation is a common weathering process in moist climates and occurs both above and below ground. When silicate minerals are hydrated by water carrying small amounts of carbon dioxide, expansion occurs, especially at sharp edges and corners, leading to a general rounding off of rocks.

exosphere The almost airless highest part of the thermosphere, and so of the Earth's atmosphere, where molecules escape into space.

extinction The complete dying out of one or more species of living things. The simultaneous dying out of a number of species is known as a mass extinction.

extrusive rocks Volcanic rocks. These are igneous rocks formed on the Earth's surface from underground magma that escaped as lava and then cooled and hardened. *See* intrusive rocks.

eye The calm center of a fierce tropical storm such as a hurricane or typhoon.

fabric The texture and structure of rock.

facies Features specific to a rock, such as the overall appearance, composition, and conditions of formation, and variations in these features occurring over a geographic area. Rock facies do not include changes resulting from weathering, metamorphism, or structural disturbance. The term is used mainly in connection with sedimentary rock, and refers to features such as grain size, bedding characteristics, and fossil content (biofacies). In relation to metamorphic rocks, facies may involve the presence of minerals, indicating the degree of metamorphic change. The term facies is also used to refer to groups of rocks thought to have been formed under similar conditions.

fall *See* autumn.

fallout The deposit of radioactive particles produced by a nuclear explosion and carried by prevailing winds in the atmosphere, often a long distance from the site of origin. Fallout enters the food chain, and isotopes with a long half-life, such as strontium-90 and iodine-131, will have effects for many years.

family A classification group consisting of a number of closely related genera.

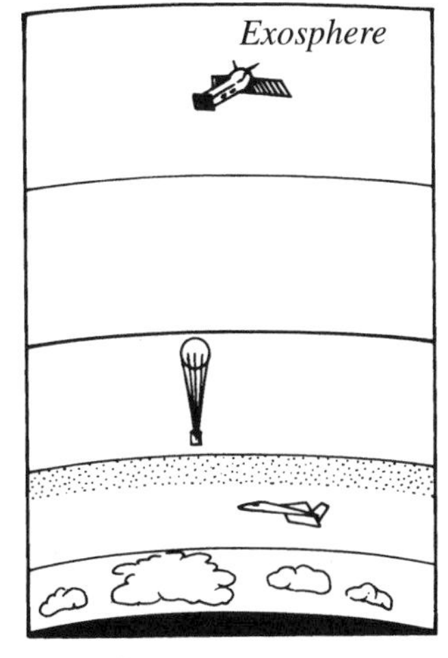

Exosphere

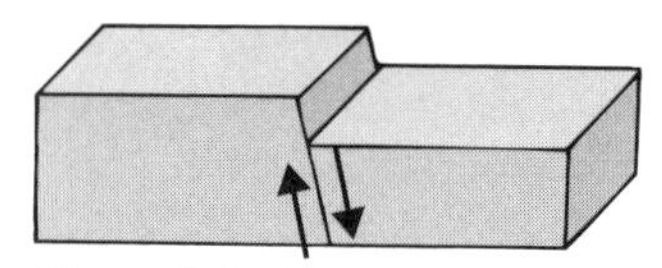

Normal fault

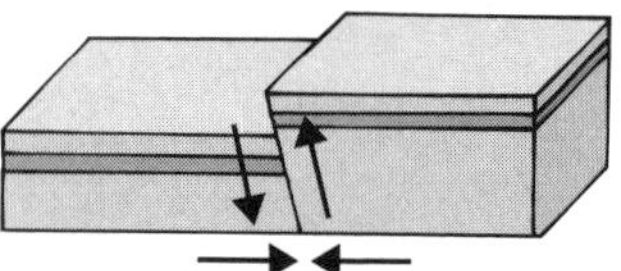

Reverse fault

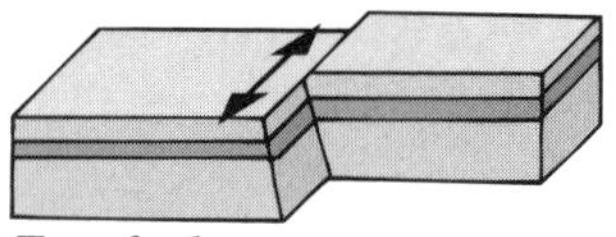

Tear fault

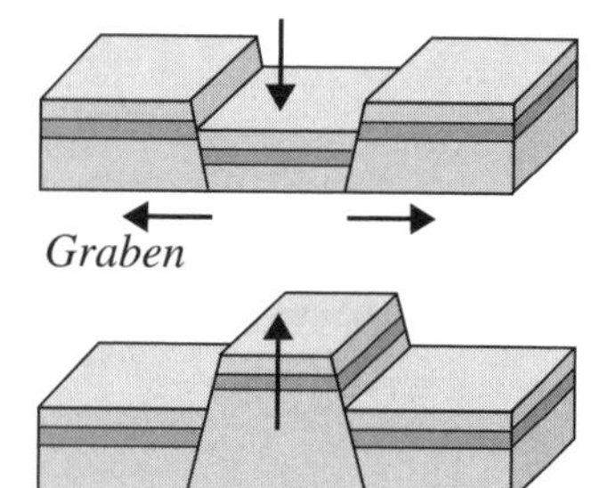

Graben

Horst

Fault

fault A fracture zone where one rock mass has moved against another in the continental and oceanic crust. Faults may extend for hundreds of miles in length. Strike-slip faults are those involving purely horizontal movements between adjacent blocks. Dip-slip faults involve directly up or down relative movement. Oblique-slip faults involve movement in directions between horizontal strike and dip slip.

fault-block A slab of the Earth's crust between parallel faults. If it stands above or below the land on each side, the slab forms either a horst or a rift valley. *See also* graben.

feldspar Any of a group of hard rock-forming minerals containing aluminum silicates of potassium, sodium, calcium, or barium. Feldspars comprise nearly two-thirds of the Earth's crust and are the principal constituents of igneous rocks. The group includes orthoclase, microcline, and the plagioclase minerals.

felsic Quartz and feldspars as a group of pale silicate minerals.

felsite An igneous rock made almost entirely of feldspar and quartz.

fermentation A enzymatic respiratory process in yeasts and other microorganisms in which glucose is broken down to pyruvate (glycolysis) and, in the relative absence of oxygen, pyruvate is converted to ethanol (ethyl alcohol) and carbon dioxide.

Ferrel cell A pattern of mid-latitude atmospheric circulation suggested by American meteorologist William Ferrel in 1856. *See also* Hadley cell.

ferromagnesian minerals Silicate minerals containing substantial proportions of iron and magnesium. They include olivines, hornblende, augite, hypersthene, and biotite mica.

finger lake A long, narrow lake occupying a valley widened and deepened by a glacier and dammed by an end moraine.

firn Old compressed snow in a glacier that has survived a summer melting season before it is converted into ice.

firn field An area of old, compressed snow (called firn or névé) accumulating at the head of a glacier. In time, compaction turns firn to ice, which resupplies the glacier as it flows slowly downhill.

fjord A long, narrow, steep-sided sea inlet invading a glaciated valley.

flint A hard, impure, opaque, brittle, even-grained, microcrystalline form of quartz (silica) found as nodules and thin layers in chalk.

flood Water drowning land that is not normally submerged. Common causes of floods are rivers swollen by rainfall and low shores swamped by storm winds that coincide with extra high tides.

floodplain Low land flanking a river and flooded when the river overflows. Sediments dumped by the river help to make floodplain soils fertile.

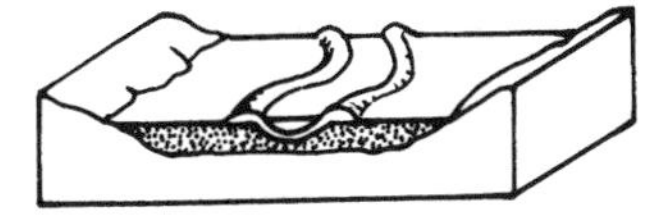

Floodplain

flow Movement of water, ice, air, or molten rock. Depending on the substance and situation, flow can be triggered by gravity, wind, or local differences in temperature, pressure, and density.

flow banding A rock structure that often develops in acidic lavas, indicating differences in composition or in the concentration of gas bubbles.

fluorocarbons Compounds formed when the hydrogen atoms of hydrocarbons are replaced by fluorine atoms. Such compounds are very stable and inert and have many functions, including the destruction of the ozone layer in the stratosphere. *See also* chlorofluorocarbons (CFCs).

fluvial Describing processes relating to flowing river and stream water.

fluvial sediments Sedimentary deposits formed on land from the sediment carried by rivers. Such deposits are often short-lived and are re-eroded as the river continues to flow, wearing away its bed and banks. Fluvial sediments include deposits of sand, gravel, metal ores, placer gold, and diamonds.

flysch facies A thick succession of redeposited marine clastic material, formed on continental margins by plate convergence prior to major orogenic uplifts.

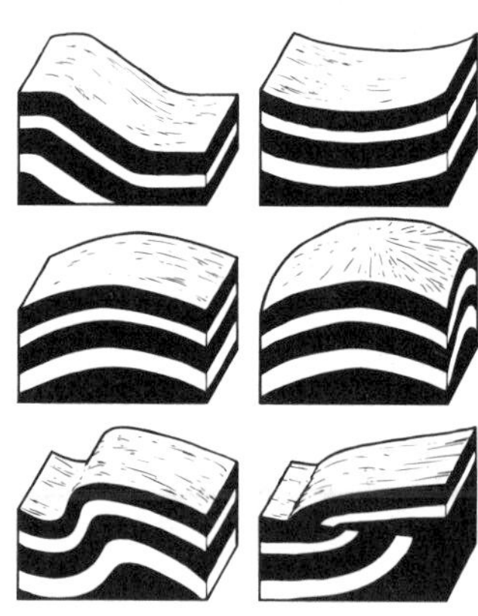
Fold

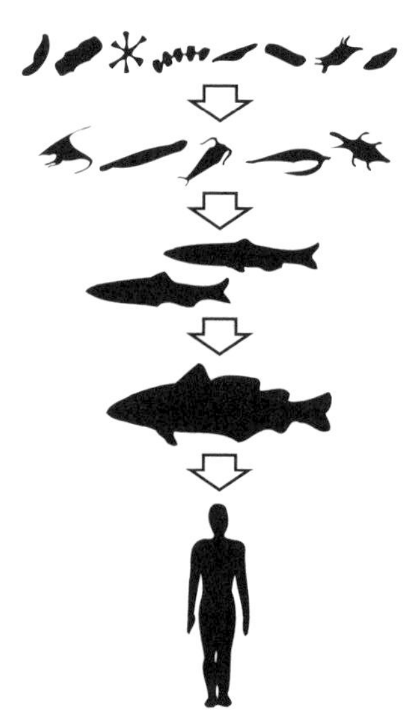
Food chain

focus In seismology, the point in the Earth's crust where an earthquake originates.

fog A cloud near the ground or over water.

fold A bend in rock layers, formed when pressure has made them plastic.

folding A change in the amount of dip of a surface relative to its original surface of deposition (bedding plane).

foliation Parallel alignment of flaky minerals in rocks that have been subjected to high pressure. *See also* exfoliation.

food chain A series of different life-forms linked by what they eat and what eats them.

food web A mesh of interlinked food chains.

foredeep A long basin filled with sediment eroded from a nearby mountain system.

foreland (1) A piece of land jutting out into the sea.
(2) Land adjoining a mountain range thrown up by the crumpling of rocks.

forest An area of trees larger than a wood. Different climatic regions favor forests of different types of tree, for instance coniferous, broad-leaved deciduous, and broad-leaved tropical.

formation A locally widespread rock type having a recognized set of characteristics and bearing a specific group of fossils.

fossil The remains or trace of a prehistoric plant or animal, organism, footprint, or dropping, preserved in sedimentary rock. *See also* living fossil.

fossil fuels Coal, oil (petroleum), and natural gas: carbon-rich substances formed from prehistoric organisms buried under the sea and subjected to pressure. *See also* hydrocarbons.

fossilization The process of forming a fossil. This can involve reinforcing, as in the case of a bone or leaf becoming impregnated with mineral deposits, or the original object can be replaced with a mold.

fracture A breakage in a rock that does not follow its natural cleavage direction or correspond to a recognizable structural plane. *See also* conchoidal fracture.

fringing reef A coral reef attached to and fringing a coast, with no lagoon between it and the land.

front The sloping boundary between masses of cold and warm air and hence of air of different density. A front is a long, sloping zone in the troposphere, featuring large changes of temperature and wind speed. So the movement of a front past a fixed location causes a rapid change in the weather at that point, with sudden changes in temperature and wind force.

frontal analysis *See* air mass analysis.

frost (1) Subfreezing conditions that can kill certain plants. (2) Frozen particles of moisture on the ground, trees, etc.

frost action The mechanical weathering process caused by the freezing and thawing of water in pores, cracks, and other openings in rocks, mainly near the surface. Water expands when it freezes, so a water-filled crack is exposed to considerable expansive forces as a result of frost. This results in fractures and flaking in rocks and in the movement and stirring of soil. A freezing cycle can cause considerable erosive action in a single night.

Frost action

fumarole A small hole in the ground, emitting steam and other hot volcanic gases.

gabbro A dense, dark, coarse-grained intrusive rock largely consisting of plagioclase feldspar and pyroxene; the intrusive equivalent of basalt.

galaxy A grouping of billions of stars and interstellar material in space. Gravity holds it together.

gale A very strong wind, registering force 8 or more on the Beaufort scale.

gangue Valueless minerals in ore.

gangue minerals Those parts of an ore deposit that are of no commercial value but cannot be avoided in mining and must be separated from the wanted ores.

gas Natural gas is a fuel consisting of methane and other hydrocarbons. It forms from the decay of organic substances and occurs in rocks underground where it may overlie petroleum.

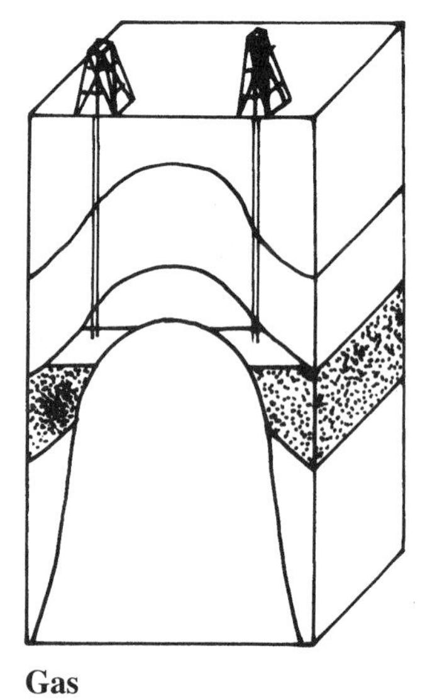

Gas

Gastropoda The snails, the largest class in the phylum Mollusca, numbering some 74,000 species.

gaussmeter An instrument for measuring the strength of a magnetic field.

gemstone A mineral valued for its beauty, durability, scarcity, and suitability for cutting into a gem.

geochemistry The study of the substances in the Earth, their distribution, and their chemical changes. Geochemical techniques for analysis include mass spectrometry, neutron activation analysis, and ion microprobe analysis. The use of such instruments has resulted in greatly improved sensitivity and accuracy of rock and soil analysis.

geochronology The scientific dating of rock formation and other events in Earth history.

geochronometry The study of the absolute age of the Earth's rocks, using radioactive decay methods. Geochronometry is used to date important events in geology, including major events in the formation of the Earth's crust, to assess the rate of evolution of organisms, and to time the advance and retreat of glaciers.

geode A hollow nodule, usually of dense chalcedonic silica rock, lined inside with quartz, calcite, or dolomite crystals. Geodes are found most often in limestone beds or less often in some shales. Many are full of water. Some have alternating layers of silica and calcite and almost show some degree of banding, implying sequences of precipitation.

geodesy The science of surveying and recording the details of the Earth's surface, the size and shape of the Earth, and its gravitational field. Geodesy has been much aided by technological advances, especially the use of surveying from artificial satellites.

geoelectricity (terrestrial electricity) The electrical and electromagnetic phenomena associated with the Earth. Natural Earth electricity results from changes in the Earth's magnetic field that induce a changing electric field and electric currents in the Earth, as a conducting medium.

geography The study of the Earth's surface features. Human geography deals with the relationship between these features and people.

geoid The Earth's theoretical form, affected by local variations in density.

geologic map A map showing solid geology (underlying rocks) and/or drift (deposits laid down by rivers, glaciers, etc.).

geologic time scale The period from the formation of the Earth to the present. The time scale is divided into four eras:
the Precambrian (4.5 billion years to 570 million years ago);
the Paleozoic (570 million years to 230 million years ago);
the Mesozoic (230 million years to 65 million years ago);
the Cenozoic (65 million years ago to the present).

geology The study of the Earth, especially its rocks and minerals and their development. Geology contains a number of separate disciplines and overlaps other sciences, including geodesy, hydrology, oceanography, meteorology, and paleontology. Geology is often subdivided into physical geology and historical geology. Physical geology deals with the composition of the Earth and the physical changes occurring on it. Historical geology is concerned with stratigraphy (the slow succession of crust-modifying events through the ages and the determination of their absolute dates) and paleontology (the study of the distribution in time of fossils of plants and animals, as determined by their location in the depth of statified rock).

geomagnetic field *See* magnetic field.

geomagnetic polarity *See* polarity.

geomagnetism The study of the Earth's magnetic field.

geomorphology The study of the origin and evolution of the shape and structure of the landforms of the Earth. Also known as rock control. Geomorphology deals with the processes of weathering, transportation, faulting, and erosion, and phenomena such as slopes, drainage patterns, and coastlines. Landforms are influenced by the differences in Earth materials at different sites and in different parts of the same site. The distribution of such materials is not homogeneous. Even in a small local area materials differ from each other in chemical composition, physical properties, and mechanical behavior. These differences are the cause of the great variety in landforms. Geomorphology is also concerned with the large-scale changes brought about by the movements of the plates of the Earth's crust (plate tectonics).

geophysical exploration Determining the physical properties of the Earth to a range of depths, usually with the intention of exploiting the findings for commercial purposes, such as the acquisition of oil, gas, water, metals, or other minerals, or as a preliminary to major engineering works, such as the construction of roads, bridges, large buildings, dams, tunnels, and so on. Geophysical exploration is also involved in archaeological work.

geophysics The study of the physical properties of the Earth. Geophysics includes such subjects as geothermometry, seismology, glaciology, oceanography, geomagnetism, and geochronology. Geophysics is considered by some geologists to be a branch of geology, but many geophysicists believe it to be a discipline of equal rank with geology.

geostationary orbit The orbit of an artificial satellite with a speed and trajectory that keep it permanently above the same point on the Earth's surface.

geosyncline A great downfold in rocks, producing an immense basin in the Earth's crust. In time, the basin may fill with sediments washed off nearby continents.

geothermal energy Useful heat obtained by pumping water past hot underground rocks.

geyser A periodic fountain forced up through a hole in the ground by the pressure of steam produced by hot rocks heating underground water.

gibbous Describing the Moon seen by reflected sunlight when this reveals more than half the Moon's disk but not all of it.

glacial geology The scientific study of the modifying effect of glaciers on land areas, on oceans, and on local and world-wide climate. Glacial geology is concerned with the theories of the origin of glaciers and glacial ages, the timing and extent of past glaciations, and the erosive and sculpturing effect of glaciation. Glacial geology is a distinct discipline from glaciology, the study of glaciers themselves. *See* glaciation; glacier; glaciology.

glacial phase One of the periods of intense cold during an ice age.

glacial snout The leading edge of a glacier. According to climatic conditions, it can be advancing, retreating, or stationary.

glaciation (1) The effects on land of ice sheets or glaciers that erode rocks and deposit the rock debris.
(2) A time when ice sheets develop and spread.

glacier A naturally accumulating mass of ice that moves downwards in the process of discharging from a head or center to its terminal dissipation zone. Glaciers originate in areas above the average snow line, which on the glacier surface is called the névé line or firn line. The most active glaciers occur in areas of heaviest snowfall, such as the maritime flanks of high coastal mountain ranges. They are composed of three substances – snow, firn, and ice. Firn is compacted snow that has not yet changed to glacier ice. The principal constituent of all the glaciers is glacier ice, containing air pockets and entrapped bubbles of water.

glaciology The study of all aspects of contemporary or modern glaciers – their characteristics and processes, mode of formation, movement, and effects. *See also* glacial geology.

glass Any vitreous extrusive igneous rock, formed from lava that has cooled so quickly that crystals have not had time to form. Glasses are unstable and over the course of millions of years undergo the formation of fine slender crystals (microlites), growing from within outward. This is called devitrification.

globe A term sometimes used for the Earth but usually describing a model of the Earth.

globular cluster A concentration of old stars; a globular cluster 100 light-years across can contain millions.

GMT *See* Greenwich Mean Time.

gneiss A general term for coarse-grained metamorphic rock with irregular granular bands of quartz and feldspar alternating with thin undulating bands of micas and amphiboles.

Gondwana or **Gondwanaland** A prehistoric southern supercontinent that included what are now South America, Africa, India, Australia, and Antarctica. *See also* Laurasia; Pangaea.

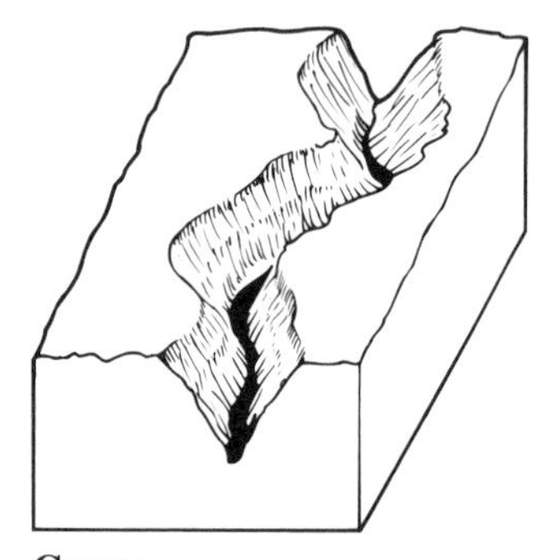

Gorge

gorge A deep, narrow, steep-sided valley, formed where a river erodes the floor of the valley faster than the sides.

gours Calcite ridges deposited where carbonate-rich water flows over a rough surface. On some limestone hillsides, gours form a terraced series of shallow basins.

graben A long, relatively narrow area of rock that has slid down between parallel faults. Unless erosion wears these down to its level, the result is a rift valley.

grade A term used to refer to the degree of metamorphism in any area. The term is also used for the fraction of a sediment falling within a particular particle size.

graded profile The profile from source to mouth of a river depositing sediment at the same rate as the river erodes it.

grain A mineral or rock fragment of sand size produced by weathering or erosion. The term is also used to refer to the plane in which a rock is most easily split.

granite An intrusive igneous rock rich in quartz and feldspar, often with mica or hornblende. Granite is hard and erosion-resistant, and granite batholiths occur in some upland regions. Granites contain at least 10% of visible quartz. They consist entirely of interlocking crystals of quartz, feldspars, biotite mica, muscovite mica, and hornblende and range in color from pale gray to pink or red. Accessory minerals such as tourmaline, apatite, sphene, pyrite, magnetite, and zircon may also be present. Most granites are very durable.

granodiorite granites Granites containing only plagioclase feldspar or a very high percentage of alkali feldspar (over 66%).

grassland Open countryside where the chief vegetation is grass. Natural grasslands include the tropical savannas and the prairies, steppes, and pampas of temperate regions.

gravel Mineral particles larger than sand grains that have resulted from natural erosion. The term is favored by British geologists and is equivalent to the American term "pebble." Gravel particles are from 2–60 mm across.

gravimetric prospecting Searching for underground oil and useful minerals with gravity meters (gravimeters): instruments that are sensitive to local differences in the Earth's gravitational field.

gravitation The universal force of attraction between two masses, such as those of the Sun and the Earth, and the Earth and the Moon.

gravity The gravitational attraction of the Earth's mass for objects on or close to its surface. The word gravity is often also used to mean gravitation.

gravity exploration The investigation of local gravitational variations caused by variations in the Earth's density from place to place. The gravitational attractive force between two bodies is directly proportional to the product of their masses and inversely proportional to the square of the distance between them. Gravity exploration involves measuring local variations to provide information on rock and other masses in the immediate vicinity. Gravity surveys are used mainly for the determination and location of large bodies, such as oil deposits, which offer the greatest gravitational differences from the surrounding land.

great circle The circle cut on the Earth's surface by a plane that passes through the centre of the globe. Segments of great circles offer the shortest route between two points on the surface.

greenhouse effect Atmospheric warming caused by gases that act like a greenhouse roof, trapping solar heat below them. The basis of the greenhouse effect is that excess carbon dioxide, water vapor, methane, chlorofluorocarbons (CFCs), and other gases absorb long-wave infrared radiation from the Earth's surface and re-radiate it back to Earth. More penetrating, shorter wavelength radiation from the Sun can, however, readily penetrate these gases to reach the surface. Waste gases produced by excess burning of hydrocarbons in human, industrial and agricultural activity are believed by most scientists to be intensifying the natural greenhouse effect.

greenhouse gases Gases whose accumulation in the atmosphere makes the Earth warmer, including carbon dioxide, methane, chlorofluorocarbons, nitrous oxide, and low-level ozone.

greenstone Any of various fine-grained, dark green, metamorphosed igneous and sedimentary rocks. Greenstone belts occur in the ancient cores of several continents.

Greenwich Mean Time (**GMT**) Time as measured at Greenwich, a part of London, England, located on the prime meridian (0° longitude). Time in all parts of the world is calculated with reference to Greenwich Mean Time.

greisen A pale, sparkling, fine-grained rock that has been exposed to the greisening process. Greisens may also contain topaz or tourmaline.

greisening A process that can operate on granitic rock as a result of exposure to very hot gases that alter the feldspars to clumps of muscovite or to the rare mica zinnwaldite.

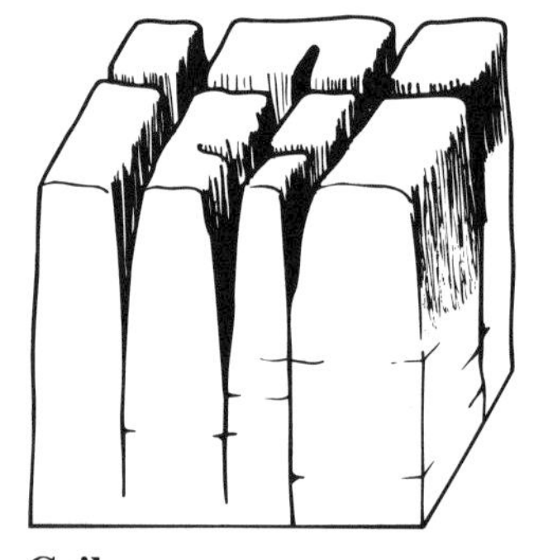
Grikes

grikes Vertical cracks up to 9 feet (3 m) deep dissecting a bare limestone surface. Grikes form where weakly acid rainwater widens joints in the limestone. *See also* clints, karst.

grit (1) A type of coarse-grained sandstone containing sharp-edged particles.
(2) Hard, sharp particles of sand.

ground moraine An undulating sheet of till dumped where a glacier or ice sheet has melted.

groundwater Water in soil and rock. It is largely rainwater that filters down through permeable and porous rock until it reaches impermeable rock. Groundwater saturating the rocks above this flows along under gravity.

groyne A wall or sturdy fence jutting out across a beach from the land. Groynes help to stop longshore drift moving sand and shingle away from the beach.

gulf A coastal inlet larger than a bay and penetrating more deeply into the land.

gully A narrow channel worn into a slope by running water, especially by runoff from heavy rainfall in semiarid regions. Gullies abound in land prone to soil erosion.

Gutenberg discontinuity The point 1,800 miles (2,900 km) deep where seismic S waves stop moving down through the Earth. Named for American seismologist Beno Gutenberg, it marks the outer rim edge of the Earth's core.

Guyot

guyot A flat-topped submarine mountain. Guyots are subsiding volcanic islands. Their summits were beveled by waves as they sank below sea level.

gypsum A mineral or rock made of calcium sulfate and water. Gypsum is an evaporite formed on the floors of seas that dried up.

hachures Short lines drawn on some maps to indicate mountains and hills. They run up and down slopes. The steeper the slope, the thicker or closer together the hachures appear.

Hadean eon The first eon (4.6–4.0 billion years ago) in the Earth's history. The Earth's surface was then hot and subjected to heavy bombardment by meteorites. Life as we know it had not appeared.

Hadley cell The tropical part of the Earth's atmospheric circulation. It consists of heated air rising from the Equator, spreading north and south, cooling and descending about 30° north and south of the equator, and then, as the trade winds, flowing back toward the equator.

hail Precipitation consisting of ice pellets called hailstones. Hail occurs during thunderstorms.

hailstone A pellet formed of concentric layers of ice that have accumulated from water droplets freezing inside a cumulonimbus cloud.

halide A chemical compound in which a halogen element (such as iodine) is linked with another element.

halite Rock salt: crystalline sodium chloride that accumulated as an evaporite on the floor of a sea or salt lake that dried up.

hanging valley (1) A tributary valley ending abruptly above a cliff or very steep slope at one side of a main valley, widened and deepened by glacial erosion.
(2) A valley ending above a sea cliff where sea is eroding the land.

hardness The characteristic of a mineral assessed by its ability to scratch another mineral or substance. The Mohs' scale of hardness is almost universally used, and is based on the principle that a hard mineral can scratch one that is softer.

headland A cape or promontory jutting into a sea, especially one bounded by cliffs.

headstream A stream forming the source of a river.

headwall The wall of rock above and behind a cirque.

headward erosion Erosion where a stream's source eats back uphill into a hillside. The cause may be a spring feeding the stream and undermining the slope above it until part of it collapses.

heath Open, uncultivated land covered mainly in grasses and shrubby plants such as heather and gorse. Heaths occur on poor soils in temperate regions.

heavy metals High-density metals such as cadmium, lead, and mercury. Heavy metals are poisonous, and careless dumping of heavy-metal wastes can create local health hazards.

hematite Ferric iron oxide; the world's chief source of iron ore.

Hemiptera An order of insects that comprises the true bugs. They include the bedbugs, plant bugs, stink bugs, assassin bugs, kissing bugs, squash bugs, lace bugs, and back swimmers. There are about 25,000 different species.

herbicides Chemicals used to kill vegetation, either selectively or indifferently. Many of them are halogenated hydrocarbons such as the chlorinated biphenyls.

Hertzsprung–Russell diagram (**H–R diagram**) A graph, devised between 1911 and 1913, that plots the colors of stars against their brightness in descending order, with the Sun midway in the sequence. It is named after astronomers Ejnar Hertzsprung and Henry Norris Russell.

hiatus Rock strata that show a break in the sedimentary succession of the geological record.

high An area of high atmospheric pressure. *See* anticyclone.

hill A raised, rounded area of land lower than a mountain.

Hirudinea The class of annelid worms known as the leeches. They are bloodsucking aquatic parasites.

hogback A long, narrow ridge formed from very steep or vertical rock layers.

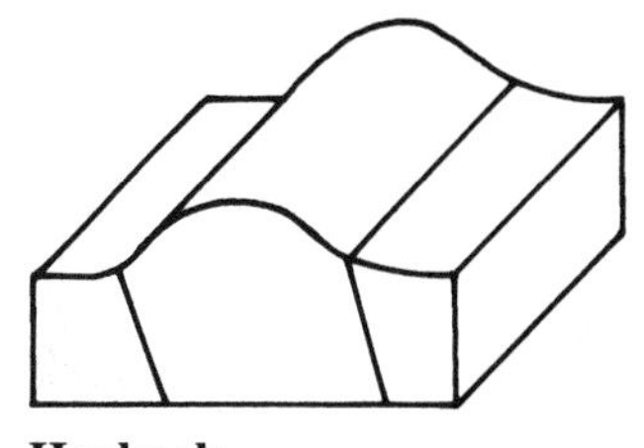
Hogback

Holocene epoch The second part, from 10,000 years ago to the present, of the Quaternary period. Holocene (recent) time covers the climatic phase since the melting of the north's Pleistocene ice sheets.

homeostasis The maintenance of constant conditions in a biological or electronic system by automatic feedback mechanisms that counter trends away from the designated fixed limits of normality.

hominid Any organism of human-like general characteristics.

Homoptera An order of sucking insects, allied to the Hemiptera, that includes aphids, leaf-hoppers, lantern flies, harvest flies, spittle bugs, and cicadas.

horizon (1) The line where sky and land or sky and sea seem to meet. The higher your viewpoint, the further away the horizon appears.

(2) A layer of soil. Many soils have three horizons: the topsoil or A-horizon, the subsoil or B-horizon, and the C-horizon of broken bedrock mingled with clay and sand. *See* subsoil; soil; regolith; topsoil.

horn A steep-sided pyramidal mountain peak formed by the backward erosion of the headwalls of several cirques. The Matterhorn is a world-famous example. *See also* arête.

Horn

hornblende Calcium-rich silicates: a group of amphibole minerals, common in metamorphic and igneous rocks.

hornfels A granular, fine-grained, flinty type of metamorphic rock produced from other rocks subjected to very high temperatures under low pressure.

horse latitudes Subtropical belts of atmospheric high pressure; calm regions of descending air located between the trade winds and westerlies.

horst A high block of land between parallel faults, usually long as compared to its width, caused by the block having risen relative to the land on either side. Horsts may have lengths and upward displacements of only a few inches, though there are those that are miles long with upward displacements of thousands of feet.

hot spot A small area of the Earth's crust heated by a mantle plume rising below it. Molten rock may emerge as a volcano.

hot spring A continuous flow of hot water from a hole in the ground. The water is warmed by hot underground rocks. Most hot springs occur near volcanoes, some of which are no longer active.

H-R diagram *See* Hertzsprung-Russell diagram.

humidity The (variable) amount of water vapor in the atmosphere.

humus The dark upper part of soil, made of the remains of dead plants, organisms, and animal remains and droppings that have lost their original structure and undergone decomposition. Humus is a stabilizing material that confers physical, chemical, and biological benefits on soils. Soils rich in humus are fertile.

Humus

Humus improves the texture, helps to bind sandy soils together into stable crumbs, loosens and aerates hard clayey soils, and increases their permeability and moisture-holding capacity. Humus helps soils to resist erosion.

hurricane A tropical cyclone tracking across the southwest North Atlantic and the Caribbean Sea. Hurricane-force winds register 12 on the Beaufort scale.

hydrocarbons Chemical compounds that contain carbon and hydrogen only. They form the fossil fuels, coal, oil, and natural gas. There is a vast number of hydrocarbons and they can be divided into four main groups: alkanes (the paraffins, such as ethane, methane, and butane), alkenes (e.g., ethylene, propene, butene), alkynes (e.g., acetylene), and arenes (e.g., benzene, toluene, and naphthalene).

hydroelectricity Electricity generated by turbines built into dams or barrages and driven by falling water.

Hydroida An order of coelenterates that includes many of the smaller jellyfish and the freshwater hydras.

hydrologic cycle *See* water cycle.

hydrology The scientific study of the Earth's water: its distribution, circulation, chemical and physical properties, and its reaction with the environment, including living things. Water, whether liquid or solid, covers about 74% of the Earth's crust. Solar energy turns a proportion of it to an atmospheric gas, which circulates widely, condenses, and precipitates, and gravity returns it to the land and the seas. This is the hydrologic or water cycle, an important aspect of hydrology.

hydrosphere The film of water coating the Earth, including water vapor, clouds, rain, snow, mist, fog, oceans, rivers, lakes, ice sheets, and the water in soil and rocks. The hydrosphere consists of the oceans, inland seas, freshwater lakes, saline lakes and rivers, soil moisture and water above the water table (vadose water), groundwater to depths of 13,100 feet (4,000 m), ice caps and glaciers. Some 97% of the weight of the hydrosphere is in the oceans.

hydrothermal deposits Minerals deposited from hot, water-rich solutions, chiefly underground and on submarine spreading ridges. Metallic mineral ores, largely formed as hydrothermal deposits.

hygrometer An instrument measuring the atmosphere's relative humidity: the amount of water vapor compared with that needed to saturate the air at the same temperature. The various types of hygrometers involve hairs, chemicals, and thermometers. *See* wet and dry bulb thermometer.

Hymenoptera The large order of insects that contains the bees, wasps, ants, and sawflies. There are probably more than 100,000 species of insect in this order.

hypabyssal Of medium-grained intrusive igneous rock that has crystallized at shallow depths below the Earth's surface.

hypabyssal intrusion An intrusion just under the Earth's surface formed from magma that failed to reach the surface. A dyke is a hypabyssal intrusion.

hypermelanic Describing igneous rocks composed almost entirely of dark minerals such as olivine and pyroxene.

ice Frozen, solidified water. Fresh water freezes at 32°F (0°C). Seawater freezes at lower temperatures, which vary with salt content. On freezing, water expands and floats on the denser liquid water below.

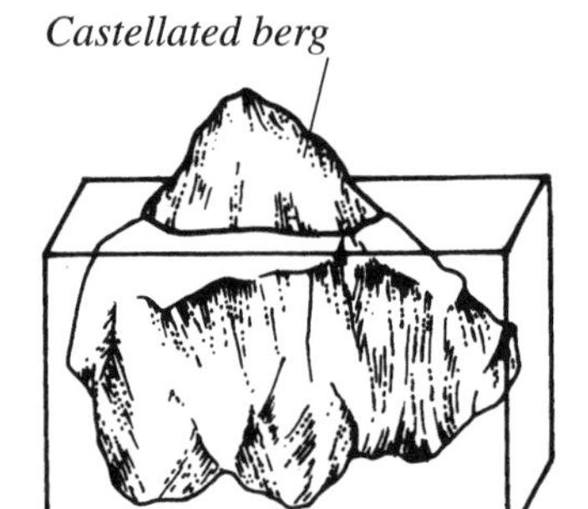

Iceberg

iceberg A floating mass of ice that has broken off the end of a glacier and fallen into the sea.

ice age One of many periods when ice sheets covered much of the Earth's high latitude regions. We live in an interglacial phase of one of several ice ages that have occurred through Earth history.

ice cap A permanent ice mass smaller than an ice sheet. There are examples in Iceland and Norway.

ice cave (1) A cave carved by meltwater in a glacier or ice sheet. (2) A rock cave containing permanent ice.

ice fall A mass of pinnacles and deep gullies in the ice of a valley glacier where its slope steepens, causing crevasses to intersect.

ice lens A small area of ice formed from water freezing in periglacial soil. The resulting expansion heaves up the overlying soil and may result in a pingo.

ice sheet An immense mass of ice covering a large land area. Ice sheets cover much of Antarctica and Greenland.

ice shelf A thick shelf of ice attached to land but floating on the sea. The ice shelves fringing Antarctica are extensions of the Antarctic ice sheet.

ice wedge A wedge-shaped mass of ice up to 100 feet (about 30 m) deep, projecting down into the ground of periglacial regions. Ice wedges form from summer meltwater filling cracks that gape open when the ground shrinks in winter.

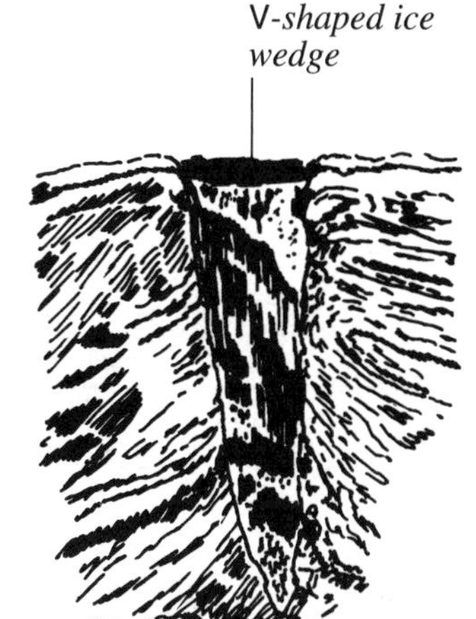

Ice wedge

igneous rock Rock formed of cooled, solidified magma. Extrusive, or volcanic, igneous rock forms as lava escapes from volcanoes and cools on the surface. Intrusive igneous rock forms from magma that cools and hardens underground.

impact crater A crater gouged in the surface of a moon or planet by the impact of a meteorite.

impermeable rock Rock that allows little or no rainwater to sink down through it.

index fossil Any type of fossil organism typical of a particular time in the Earth's past, used as an indicator of age. Index fossils help geologists to date rocks of the same age from different parts of the world.

inertia The tendency of an object to remain stationary or to move in a straight line unless affected by an external force.

Insecta One of the classes in the phylum Arthropoda and the largest of all the classes of animals. Nearly a million species have been described, but the total is known to be much larger. The Insecta are divided into many different orders.

inselberg An isolated, steep-sided hill rising abruptly from a tropical plain. Inselbergs occur in savannas as remnants of eroded rock masses or plateaus.

interglacial phase A mild or warm phase between frigid glacial phases of an ice age. During an interglacial phase, ice sheets retreat and forests may replace tundra vegetation.

intermittent stream A stream, usually in a desert region that carries water only at intervals, such as during a flash flood.

international date line An imaginary line around the world where the date changes by a day. With local deviations, it follows the 180° meridian, halfway around the world from the Greenwich meridian. Places just west of the line are nearly 24 hours ahead of places just to its east.

International date line

interplanetary matter Small solid particles and gas present in space between the planets. Most of our interplanetary gas derives from the Sun. The solid material comes from the asteroids and from comets.

interstellar matter Small solid particles and gas present between the stars. Interstellar matter makes up a small but significant proportion of the total mass of the universe. It is the raw material from which new stars are formed.

intrusive Describing a body of rock, usually igneous, that is found within preexisting rocks.

intrusive rocks Igneous rocks formed from magma that has found its way into spaces in the Earth's crust and cooled and hardened to form igneous rock. Intrusive rocks include granite. *See* extrusive rocks.

ion An atom with a positive or negative electrical charge because it has, respectively, lost or gained electrons.

ionic bonding Chemical bonding in which electrons form ions by moving from one atom to another.

ionosphere An ion-rich upper layer of the Earth's atmosphere where the auroras occur. Its ions (electrically charged atoms) are produced by solar ultraviolet radiation and radiation from space. The ionosphere, which is less rigidly layered than was formerly believed, reflects certain radio waves back to Earth. *See also* mesosphere.

iron A silvery metal forming 5% of the Earth's crust, where it is chemically combined with oxygen, carbon, and certain other elements in ores including hematite and magnetite.

irrigation Watering land to promote crop growth. This can involve river valleys dammed to create reservoirs, canals and tunnels, and ditches for transporting water and huge mobile sprinklers.

island A piece of land completely surrounded by a river, lake, sea, or ocean.

island arc A curved string of volcanic islands on the continental side of an oceanic trench.

isobar On a climatic map or weather chart, a line passing through places that have the same atmospheric pressure at the same time of day or year. On such maps or charts, isobars show the distribution of areas of high and low pressure. Isobars are important in weather forecasting and meteorology. Roughly circular sets of closely placed isobars around centers of low pressure at mean sea level indicate areas of bad or unsettled weather. Closed isobars around high-pressure centers suggest generally good weather.

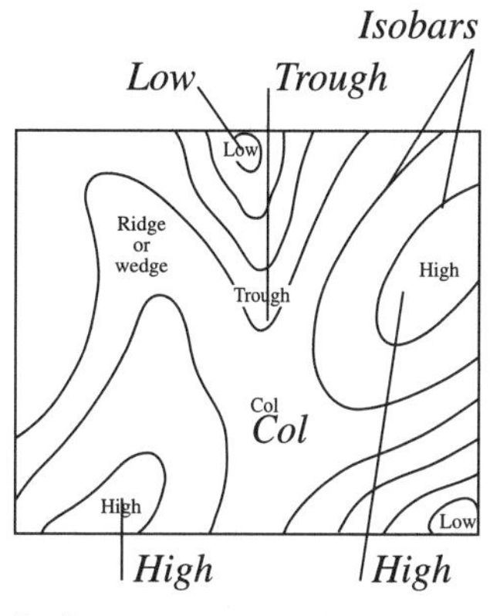

Isobar

isohyet On a map, a line passing through places that have the same rainfall over the same period.

isoptera The order of insects containing the termites.

isostasy The state of balance of the Earth's crust as it floats on the denser mantle. Mountains are balanced by deep roots of crustal rock.

isotherm On a map, a line passing through places that have the same air temperature at the same time of day or year. Isotherms are used to show the distribution of areas of relatively high and low temperature.

isotopes Atoms of the same element sharing the same atomic number (number of protons in the nucleus) but differing in atomic mass (number of protons plus neutrons). Because they have the same number of protons, they also have the same number of electrons and thus the same chemical properties. There are, however, differences in their physical properties. Many isotopes are radioactive and can be used for labeling atoms to investigate biochemical reactions. They can also be used in radioactive dating techniques. *See* radiocarbon dating; radiometric dating.

isthmus A narrow neck of land joining two larger land areas.

jasper An impure form of quartz, variously colored and banded and of variable hardness, used as a semiprecious gemstone.

jet A hard, black mineral, a form of lignite coal that takes a high polish and can be carved and turned to make beads and other decorative items.

jet stream A high-altitude, high-speed westerly wind that blows around the world. A jet stream is a narrow, fast-moving current of wind flanked by more slowly moving currents. Jet streams occur mainly in the zone of prevailing westerlies above the lower troposphere, the two major jet streams being the polar front jet stream and the subtropical jet stream. They may have serious implications for aircraft operations because of the high speed of the wind at the jet core and the rapid variations in wind speed in the area around the stream.

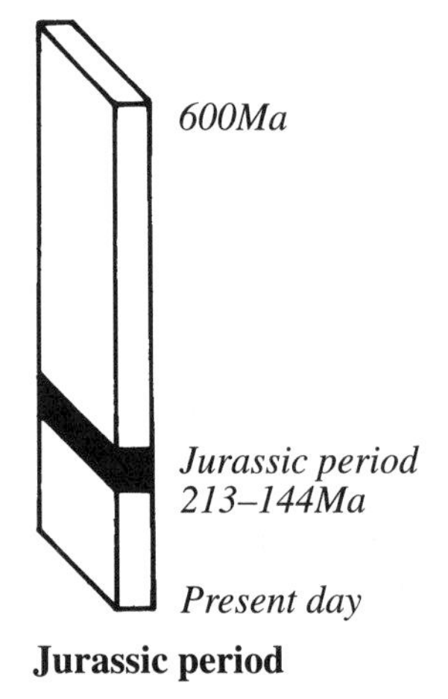

Jurassic period

joint A crack in rock, formed along a line of weakness.

Jupiter The largest planet in the solar system, with a diameter more than 11 times that of the Earth. It consists mostly of the gases hydrogen and helium.

Jurassic period The second period (213–144 million years ago [Ma]) of the Mesozoic era, a time when the largest of all dinosaurs flourished. The Jurassic period was preceded by the Triassic and followed by the Cretaceous. Jurassic plants were mainly ferns, gymnosperms, and angiosperms. Reefs were built by algae, and the most important marine invertebrates were the ammonites, of which enormous numbers of index fossils remain. The Jurassic features a great many reptiles, the dinosaurs being the dominating forms on land. The first bird, archeopteryx, arose in the late Jurassic. Almost all groups of modern fishes were present. There were also many insects, such as flies, butterflies, and moths.

Jurassic table The system, mainly of sedimentary rocks, formed during the Jurassic period. It was named after the Jura Mountains in Switzerland.

kaolinization The process by which the feldspars in granitic rock may be partially or wholly converted to fine, flaky clay minerals and fine, colorless mica called sericite.

karst Limestone landscape with a largely bare, rocky surface and rivers that flow through underground caves.

kettle hole A hollow in land covered by a glacial drift deposit. Kettle holes contained blocks of ice that melted, leaving depressions. Many are occupied by small lakes.

knickpoint A break in the slope of a stream bed, often with rapids. It usually marks the limit of headward erosion reached by a stream deepening its bed from its mouth upstream after sea level fell or land level rose. *See also* rejuvenation.

Krebs cycle *See* citric acid cycle.

K-T boundary The short geological period at the junction between the Cretaceous and the Tertiary periods, about 65 million years ago, which saw the extinction of the dinosaurs together with a major proportion of the living things on Earth. The cause of this extinction remains uncertain.

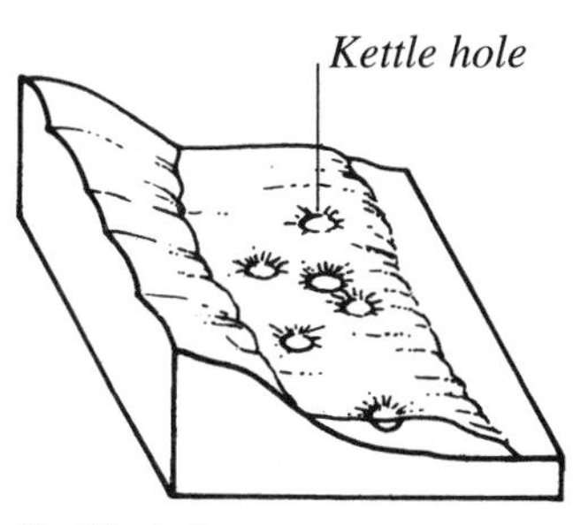

Kettle hole

laccolith A lens-shaped mass of intrusive igneous rock that pushes overlying sedimentary rocks into a dome.

lagoon A shallow area of water partly or wholly cut off from the sea by a low-lying strip of sand, shingle, or coral that forms a spit, bar, or atoll.

lake A large sheet of water surrounded by land, or, more rarely, ice. So-called land-locked seas such as the Caspian Sea are really giant lakes.

landfill Disposal of hazardous or other waste by tipping it in a hole in the ground. Consequences can be an explosive buildup of underground methane gas and contamination of water supplies.

landform A distinctive natural configuration of the land surface.

landslide The sudden slide down a slope of a mass of rocks or soil. Landslides can happen when water lubricates a line of weakness in rock.

land use The economic use to which land is put. For instance, land may be classified as cropland, grassland, or forest.

lateral erosion The gnawing away of the banks of a meandering river by the flow of water. Lateral erosion chiefly occurs where a strong flow of water undermines the bank on the outer side of each bend.

lateral moraine A line of rocky rubble lying on one side of the surface of a valley glacier. It results from stones and rocks falling onto the glacier from the steep slopes above. A similar lateral moraine occurs on the glacier's other side.

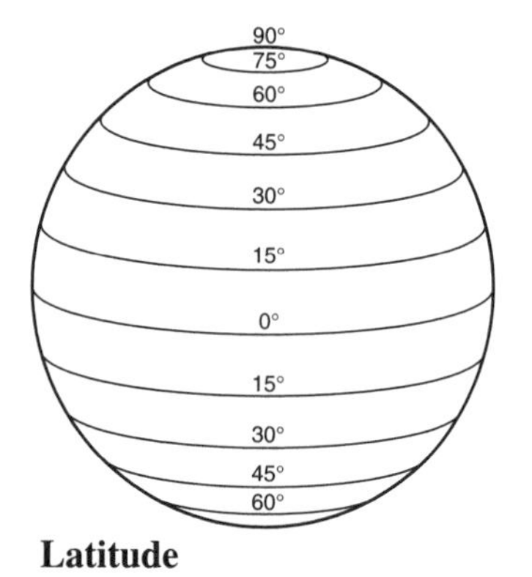

Latitude

latitude Location north or south of the equator. Latitude is measured from the center of the Earth in degrees north and south of the equator. The equator's latitude is 0°. The poles are 90° north and south of the equator.

Laurasia A prehistoric northern supercontinent formed about 200 million years ago after the global supercontinent Pangaea began breaking up. It included North America, Greenland, Europe, and much of Asia. *See also* Gondwana; Pangaea.

lava Molten rock when it appears at the Earth's surface from a volcano through vents and fissures. Silica-rich acid lava hardens before flowing far. Rapid cooling at the Earth's surface can transform fluid lava into a dense-textured volcanic rock of tiny mineral crystals or glass. Basic lava flows further before it solidifies, giving rise to coarse-grained igneous rock, such as granite or gabbro. In many volcanic eruptions lava is blown out with explosive force so that it fragments in the atmosphere. These small pieces harden rapidly and fall to Earth to form thick layers of volcanic tuff and related pyroclastic rock.

leaching (1) The removal of a soluble substance from an associated insoluble solid by dissolving it in a suitable solvent.

(2) The process by which rainwater washes soluble salts out of the upper soil into a lower soil layer.

leap second A correction to the calender necessary to keep the accepted atomic time standard in synchronization with Earth time. The leap second is needed because of slight slowing of the Earth's rotation.

Lepidoptera The order of scaly winged insects that includes the 100,000 or so species of butterflies, moths, and skippers (Hesperiidae).

leukocratic Describing igneous rocks of pale color, formed from light-colored minerals such as quartz and feldspars.

levee The naturally raised bank of a river crossing a floodplain. It consists of alluvium deposited when the river overflowed its banks. Some levees become so high that the river level between them is higher than that of the surrounding plain.

lightning Elongated sparks produced by the passage of large electric currents through the air between points of greatly differing voltage. The size of the potential difference can be estimated from the fact that for a spark to bridge one centimeter of air containing water droplets, about 10,000 volts are required. Rapidly upward-moving air in cumulonimbus clouds produces separation of electrical charges, the upper regions being positively charged and the central and lower parts mainly negative. When the potential difference reaches the breakdown point between adjacent charge centers, or between the cloud and the ground, lightning occurs.

light-year The distance traveled by light in one year: about 5,879,000,000,000 miles (9,461,000,000,000 km).

lignite *See* brown coal.

limestone Sedimentary rock made mainly of carbonates of calcium and magnesium and laid down in the sea. Limestone can occur as the skeletal remains of coral polyps or other marine organisms or in the form of a precipitate. They are the most abundant of the nonclastic rocks, and are by far the greatest reservoir of the element carbon on or near the surface of the Earth. The chief minerals of limestones are calcite, aragonite, and dolomite. Most of our knowledge of invertebrate paleontology and of the evolution of life on Earth comes from the fossils contained in limestone.

limon *See* loess.

limonite An important, earthy, yellow-brown iron ore often found in sandstones, bauxite, and weathered basalt.

lineation The parallel orientation of elongated particles on a planar surface in a metamorphic rock.

lithification Processes that change sediments into solid rocks, for instance, compaction due to compression combined with the precipitation of a mineral cement.

lithosphere The Earth's crust, both continental and oceanic, coupled to the rigid upper mantle above the asthenosphere. The lithosphere consists of numerous tectonic plates that are able to move relative to each other so as to alter the profiles of landforms. This building and altering process is called plate tectonics. The lithosphere varies in thickness from about 40 miles (60 km) under parts of the oceans to about 190 miles (300 km) in continental regions. The term *lithosphere* has been used in the past to distinguish the rocky part of the Earth from the watery part (hydrosphere), the gaseous part (atmosphere), and the living part (biosphere).

little ice age A period during which the average temperature of the Earth declines by a few degrees. The last little ice age persisted from about 1500 to 1900, when the temperature was about 3° below average.

living fossil An extant organism with features closely resembling those of a fossilized organism, such as the Gingko tree.

load Material transported by moving water, ice, or air. Part of a river's load is dissolved. Tiny particles travel suspended in the water. Heavier, larger particles slide, roll, and bounce over the stream bed.

loam Fertile soil consisting of a mixture of sand, silt, and clay. Loam holds nutrients, moisture, and air.

lode An extended vein of minerals or a system of veins. The term is sometimes limited to a productive vein.

loess A layer of gray or yellow dust windblown from a desert or the edge of an ice sheet. In places some has been redeposited by rivers. Loess is also called adobe in North America and limon in Europe.

longitude Location measured in degrees east or west of Greenwich, London, which lies on the so-called prime meridian. *See also* meridian; prime meridian.

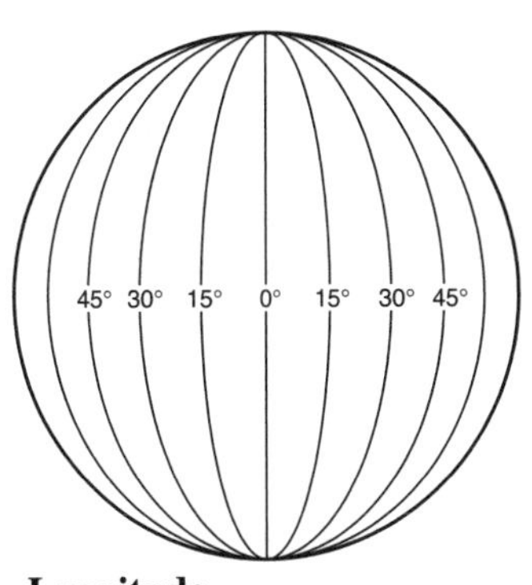

Longitude

longshore drift The movement of material along a shore. This occurs where waves break obliquely onshore, carrying sand grains up a beach at an angle. Gravity makes the water and sand slide back down to the sea. As a result the sand grains follow a zigzag path along the beach.

longwave *See* L wave.

lopolith A saucer-shaped mass of intrusive igneous rock injected between older rock layers. Some lopoliths measure many miles across.

Love wave A type of L wave caused by earthquake and polarized so that it is transmitted horizontally along the Earth's surface. A Love wave travels slightly faster than a Rayleigh wave. The wave is named for the English mathematician and physicist Augustus Edward Hugh Love.

low Any area of low atmospheric pressure, for instance a depression.

luminosity A star's essential brightness as distinct from its apparent magnitude (apparent brightness) when viewed from Earth.

L wave or **long wave** or **surface wave** An earthquake wave traveling along the Earth's surface. They arrive later than the P (primary) waves and S (secondary) waves. The two types of L wave are called Love waves and Rayleigh waves. *See* Love wave; Rayleigh wave; P waves; S waves.

Ma An abbreviation for a million years.

mafic A term describing dark, dense, rock-forming, mainly silicate minerals rich in iron and magnesium; also the igneous rocks that they form.

magma Liquid or semi-liquid molten rock beneath the Earth's surface. Magma that has cooled and hardened underground or on the surface forms igneous rock. Most magmas are melted silicates with crystals and gas, but studies of igneous rock indicate that some magmas must consist of melted carbonates, oxides, phosphates, sulfides, and sulfur. Basaltic magmas are believed to originate in the mantle. Rhyolitic magmas probably arise by crystallization of basaltic magmas or by melting of crustal rock.

magnesium A light, silvery white metallic element. It occurs in various compounds dissolved in seawater and in various silicate minerals.

magnetic exploration The mapping of local variations in magnetic field strength to determine the location, size, and shape of rocks and ores that contain minerals that have become magnetized by induction from the Earth's magnetic field. Sedimentary rock is much less susceptible to magnetic induction than igneous or metamorphic rock, so the major magnetic changes come almost wholly from the latter.

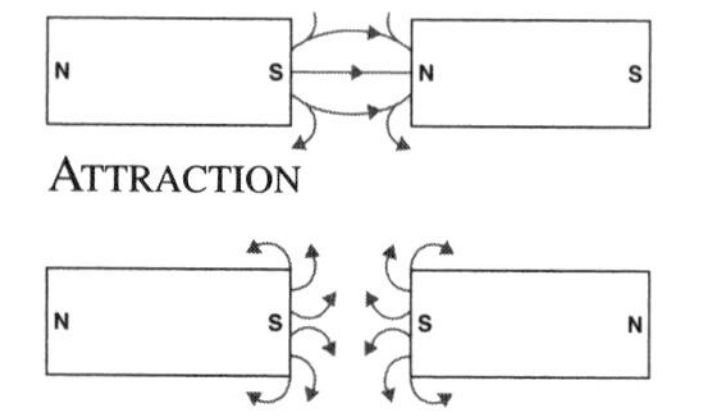

Magnetic field

magnetic field or **geomagnetic field** Imaginary flux lines (lines of magnetic force) curved around the Earth between its north and south magnetic poles. This magnetic field results from a dynamo effect created by the rotation of the Earth's crust and the mantle around the Earth's core.

magnetic poles Locations in the world's far north and far south to which a magnetic compass needle points. They do not coincide with the geographical poles, and their locations vary through time.

magnetic reversal A reversal of the Earth's magnetic field, where north and south magnetic poles switch places. Records of past magnetic reversals are alignments of magnetized particles in rocks formed at the time.

magnetometer An instrument measuring the strength of a magnetic field. Its detection of local variations in the Earth's magnetic field can help prospectors find mineral deposits.

magnetosphere A roughly doughnut-shaped region in space containing the Earth's magnetic field and magnetically trapping or deflecting charged particles. It extends 40,000 miles (64,400 km) beyond the Earth. Saturn, Jupiter, and Uranus are among other planets known to have magnetospheres.

magnitude (1) In astronomy, a star's brightness, measured as either absolute magnitude or apparent magnitude (brightness in the sky) as seen from the Earth.
(2) In geophysics, the energy released by an earthquake.

main sequence On a Hertzsprung-Russell diagram, the region where most stars (i.e. the Sun-like stars) occur. *See* Hertzsprung-Russell diagram.

manganese nodule A type of black or brown metallic lump found on parts of the deep ocean floor. Manganese nodules up to several feet across form from substances precipitated from seawater.

mantle The intermediate and most prominent layer of the Earth, a zone of dense, hot rock, 1,800 miles (2,900 km) thick, lying below the Earth's crust and above the core and extending roughly halfway to the center. Some regions of the Earth's mantle are semi-molten and flow. The mantle occupies about 84% of the Earth's volume and 68% of its mass. It is probably composed of ultramafic rocks, such as peridotite, enstatite, or eclogite.

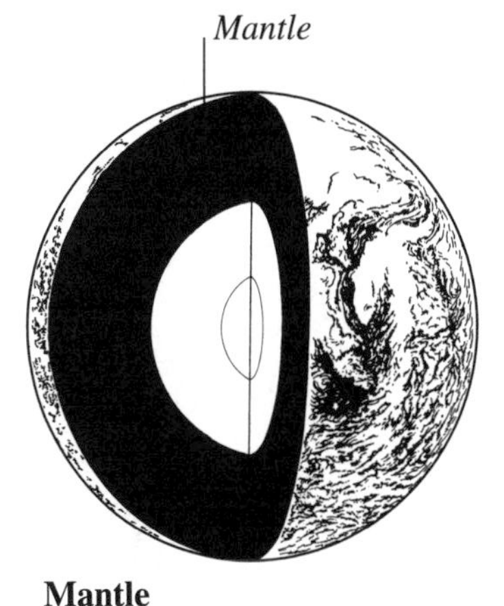

Mantle

mantle plume A plume of molten rock rising from the mantle and burning a hot spot through the Earth's crust above.

map The surface of a planet or a moon drawn to scale on a flat sheet. Small-scale maps show a large area in little detail; large-scale maps show a small area in great detail. Thematic maps show various aspects of the Earth, for instance political maps stress political boundaries, physical maps show variations in the Earth's surface, and geological maps stress rocks. *See also* topographic map.

map projections Drawings that use mathematics to show the Earth's curved surface on a flat sheet of paper. Different projections accurately show direction, shape, or area, but none shows a large region without some distortion.

marble A hard, shiny, patterned metamorphic rock formed from recrystallized limestone or dolomite.

marine sediments The accumulation of mineral and organic material that has settled from the water onto the ocean floor. Marine sediments vary greatly in composition and physical characteristics, depending on factors such as distance from land, depth of water, and differences in the sources of sediment. Sediments may be derived from land or may originate in the sea. Biogenic sediments are those formed from the skeletal remains of various kinds of marine organisms, some of which extract silica from the seawater.

marl A gray or blue-gray, crumbly, chalk-like and non-hardened calcium carbonate deposit that is formed in some freshwater lakes, partly by the action of certain aquatic plants. The clay content of marls varies from small to large. Marl will, in time, harden into marlstone, or marlite.

Mars The planet fourth in distance from the Sun and little more than half the diameter of the Earth. Mars is the planet most like the Earth, but colder, with a thin atmosphere and no liquid water.

marsh Wetland that supports sedges, rushes, or other nonwoody plants that love moisture. In many estuaries, silt trapped by salt-tolerant plants builds mudflats into salt marshes.

mass The amount of matter an object contains, indirectly measured by its inertia: its resistance to a force trying to alter its speed and direction.

massif An erosion-resistant, very large topographic or structural feature often of greater rigidity than the surrounding rock. A mass of ancient rocks forming an upland region. It can be a partly dissected plateau, with separate peaks.

massive A term applied to a rock or mineral that is either unusually heavy or that has no particular recognizable features – that is, lacks any form or structure.

mass movement Soil or rock moving downhill by gravity. Gradual movement is creep. Flowing movement produces mudflows and solifluction. Sudden movement causes landslides and rockfalls. Mass movement is often triggered by rainwater lubricating loose material on a hillside. *See also* solifluction.

matrix Any fine-grained material occurring between grains or clasts in a sedimentary rock. The interstitial material between larger crystals, particles, or fragments.

maximum thermometer A thermometer registering the highest temperature reached during a certain time. Rising mercury may push a short metal strip up a tube where it stays after the mercury sinks.

mb *See* millibar

mean (arithmetic mean) The total number of a set of items divided by the number of items.

meander A curve in a river in which the water channel swings from side to side in a series of loops.

mechanical weathering or **physical weathering** The breakup of rock at the Earth's surface caused by changes in pressure. Heating and cooling, plant roots, or the growth of salt crystals may shatter rock by widening cracks or making layers flake off. *See also* chemical weathering.

medial moraine A low ridge of rubble running down the middle of a glacier. It consists of two lateral moraines that combined where two valley glaciers converged.

melanocratic Describing dark igneous rock consisting mainly of mafic (dark) minerals.

Mercalli scale A scale devised by Italian seismologist Giuseppe Mercalli for measuring the locally felt intensity of an earthquake. The Richter scale is now more often used.

Mercator projection A cylindrical map projection first used in the 1560s by Flemish cartographer Gerardus Mercator. It shows direction accurately but greatly exaggerates area and scale in high latitudes.

Mercury The planet closest to the Sun. Its diameter is only two-fifths that of the Earth. Mercury is hot, dry, and almost airless.

meridian A line of longitude between the Earth's North Pole and South Pole. Meridians cut the equator at right angles. *See* longitude.

mesa A steep-sided plateau capped by resistant horizontal rock layers. Mesas are like buttes but larger. Both are remains of more extensive plateaus, dissected by erosion. They occur in semiarid regions. *See* butte.

mesocratic Describing igneous rocks consisting of roughly equal proportions of dark and light minerals.

mesopause A level in the ionosphere above which temperatures rise as altitude increases instead of falling as they do in the lower ionosphere.

mesosphere The Earth's atmosphere between the stratopause and mesopause. Between these, temperature falls with increasing altitude. The mesosphere overlaps the lower part of the ionosphere.

Mesozoic era The geological era between the Paleozoic and Cenozoic eras. It lasted from about 245 to 65 million years ago. The Mesozoic is sometimes known as the Age of Dinosaurs.

metal A type of element that can be shaped in certain ways and conducts heat and electricity. Useful metals such as iron and copper mostly occur chemically combined with other substances in the Earth's crust. *See* ore.

metamorphic aureole The area of rock metamorphosed by heat or pressure from the intruded mass of igneous rock that it surrounds.

metamorphic rock Sedimentary or igneous rock altered by great heat or pressure. Examples are slate (metamorphosed shale), marble (metamorphosed limestone), amphibolite (metamorphosed basalt), and schist (metamorphosed slate or basalt).

metamorphism The changes brought about in preexisting rock masses by temperature, pressure, and loss of any volatile content, but not weathering or sedimentation. Metamorphism may result in the production of new structures, or textures, or new minerals, and while conferring a distinctive new character to rock, does not involve the total loss of individuality of a rock mass. Metamorphism does not result in widespread chemical change in a rock but only the recrystallization of new individually distinct and homogeneous parts.

meteor or **shooting star** A glowing scrap of rock or metal burning up as it speeds through the Earth's atmosphere. About 100 million meteors a day enter the Earth's atmosphere. Most of them burn up before reaching the ground.

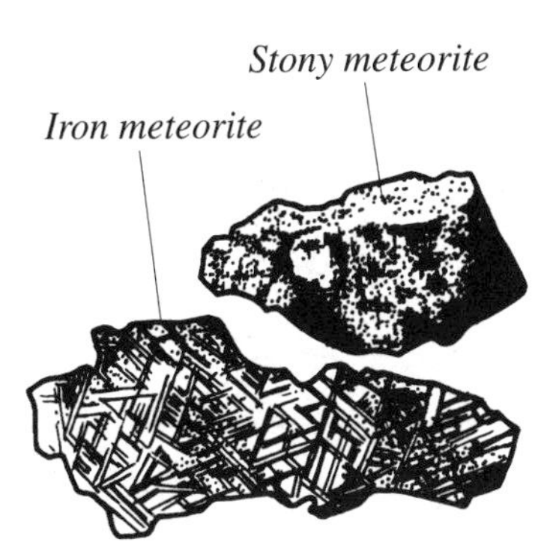

Meteorite

meteor shower A group of meteors seen from the Earth and consisting of remnants from a disintegrating comet.

meteorite A rock or metal lump that has survived the passage through the Earth's atmosphere and has fallen to Earth.

meteoroid swarm A group of cometary fragments orbiting the Sun.

meteorology The scientific study of the atmosphere, especially of changes in temperature, pressure, precipitation, and other factors influencing weather. Collecting and collating meteorological data enables scientists to make weather forecasts.

mica Any of a group of silicate minerals that split easily into thin sheets. Mica appears as shiny flakes in many igneous and metamorphic rocks. Biotite and muscovite are two forms of mica.

microcontinent A type of large island such as those that probably preceded, and formed the nucleus of, the first continents.

microlite Very small crystals, found in glassy rocks, often prism-shaped and visible only by high-power microscopy.

microplate Any small lithospheric plate with identifiable margins.

midoceanic ridge *See* spreading ridge.

Milky Way The spiral galaxy that is home to our solar system. Its 150 billion stars extend some 500,000 light years through space.

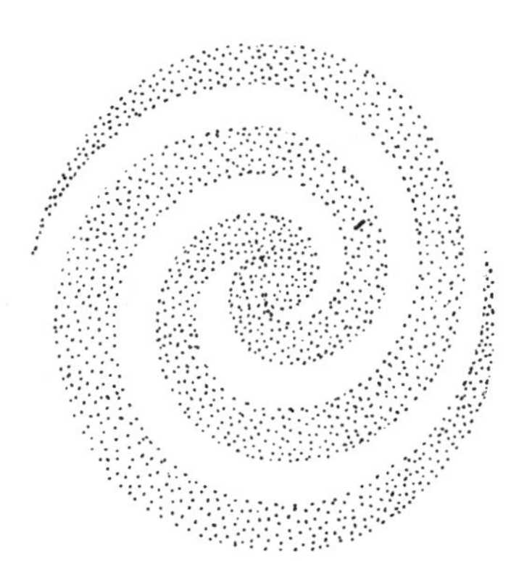

Milky Way

millibar (**mb**) A unit of atmospheric pressure used in the study of weather and climate. One-thousandth of a bar. A bar is a cgs unit of pressure equal to 10^6 dynes per square centimeter, and is equivalent to 10^5 newtons per square meter.

mineral A natural inorganic substance with distinct chemical composition and internal structure. Various kinds of minerals form the ingredients of rocks. Quartz (silicon dioxide) and feldspars (aluminum silicates) are by far the most plentiful rock-forming minerals. Coal and oil, although often loosely listed under the mineral resources of a region, are not minerals, being complex mixtures without definite chemical formulae, but metal ores are classified as minerals. Minerals may occur in pure form as individual crystals or may be widely disseminated as admixtures in rocks or other minerals.

mineral nomenclature The naming of minerals. For historical reasons, this is somewhat confused. Some names arise from the uses of minerals. Graphite, for instance, was used for writing, magnetite for attracting iron. Some names, such as rhodonite (rose color) and cryolite (ice stone), are based on appearance. One of the most plentiful minerals, quartz, has a variety of traditional names, including amethyst, agate, citrine, cairngorm, and jasper. Most modern mineral names end in "ite" but many also have a chemical name. Thus, quartz is silicon dioxide, galena is lead sulfide, zincite is zinc oxide, halite is common salt (sodium chloride), and feldspars are aluminum silicates of potassium, sodium, calcium, or barium.

mineralogy The scientific study of minerals: their form, structure, chemistry, and other properties.

minimum thermometer A thermometer registering the lowest temperature reached during a certain time. It is usually combined with a maximum thermometer.

Miocene epoch The fourth and longest epoch (about 24–5 million years ago) in the Tertiary period of the Cenozoic era. During this time mammals reached their greatest variety.

mirage An optical illusion caused by the bending of light passing between layers of air of differing density.

Mississippian period In North America, the first part of what is elsewhere called the Carboniferous period – about 360–320 million years ago. *See also* Pennsylvanian period.

mist Water droplets suspended in air and thus reducing visibility, although less severely than fog. Mist cuts visibility to about 0.6–1.2 miles (1–2 km).

mobile belt A belt of rocks formed from island arcs, submarine plateaus, etc., swept up against a preexisting microcontinent by seafloor spreading, and folded up to form a line of mountains.

Moho *See* Mohorovičić discontinuity.

Mohorovičić discontinuity (**Moho**) A level below the Earth's surface where earthquake waves suddenly change speed. Discovered by Croatian geologist Andrija Mohorovičić, it marks the boundary between the Earth's crust and the mantle. Under the oceans the Mohorovičić discontinuity is generally 6–7 miles (10–12 km) deep; under the continents it is usually 20–22 miles (33–35 km) deep.

Mohs' scale A scale showing the relative hardness of different minerals by their scratch resistance. German mineralogist Friedrich Mohs devised the scale in 1822.

molasse facies Shallow marine and nonmarine sediments formed from erosion within and around fold belts during and following their elevation into mountain ranges.

Mollusca A large phylum of the animal kingdom containing many similarly formed, soft-bodied, often shelled creatures, including the snails, slugs, oysters, clams, squids and octopuses. There are believed to be more than 110,000 different species of mollusks.

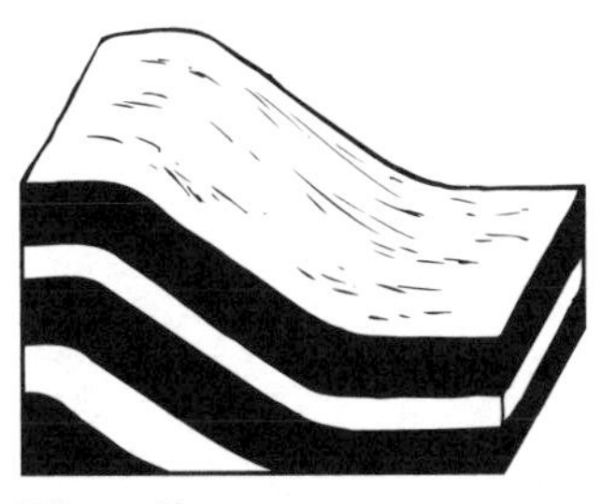

Monocline

Monera One of the five taxonomic kingdoms of living things in the five-kingdom system of classification. The Monera include blue-green algae and the bacteria.

monocline A simple fold in otherwise flat rock layers.

monoculture Cultivation of one crop to the exclusion of others. Typically it maximizes use of farm machinery but increases the risks of crop disease, pest infestation, and impaired soil structure.

monsoon A seasonally reversing wind system that affects much of tropical Asia. It blows inland in summer and offshore in winter. The summer monsoon brings rain to much of south and southeast Asia. The air of the winter monsoon is dry.

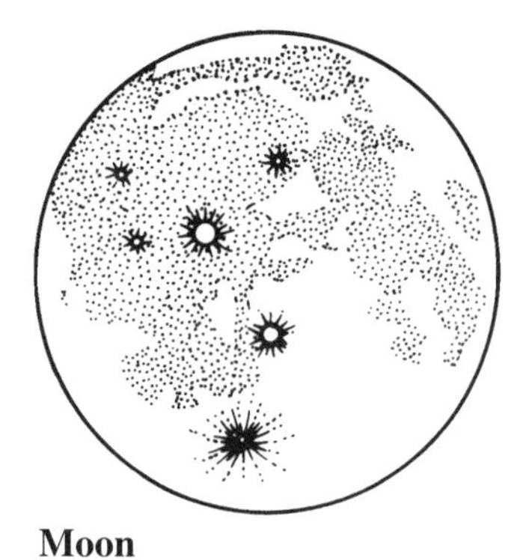

Moon

moon A heavenly body kept in orbit around a planet by gravitational force. The Earth has only one moon, but some planets have more; Saturn probably has dozens of moons.

moraine Rock debris moved by a glacier or ice sheet; also, such rock debris dumped when a glacier or ice sheet melts.

mountain A mass of land higher than a hill and standing significantly above its surroundings. Volcanic eruptions and folding, faulting, or upwarping of the Earth's crust can all produce mountains. Most occur as ranges or chains.

mountain chain A great belt of mountains, more extensive than a range. Submarine spreading ridges are the Earth's longest mountain chains.

mountain range A belt of mountains, such as the Rockies.

mud A fine-grained sediment formed from clay, silt, water, and often organic substances. Mud deposits occur on the floors of lakes and oceans.

mudflow Mud flowing swiftly down a canyon and spreading out where the canyon ends.

mudrocks or **mudstones** The finest-grained of the detrital sedimentary rocks. They contain the clay minerals that may include illite, kaolinite, smectite, chlorite, glauconite, sepiolite.

mudstones *See* mudrocks.

mutualism An association between plant and animal species that provides advantage to both. Grazing herbivores crop most forms of vegetation and, in doing so, may destroy aggressively growing plants. In cropping grasses, however, they are unable to destroy them because these plants grow from the bases of the leaves rather than from the tips. When herbivores are removed from an ecosystem, actively pioneering plants such as brambles, trees, and hardy shrubs flourish. Many associations are even more direct. Lichens, for instance, consist of a fungus incapable of photosynthesis and closely associated algae. The body filaments of the fungus (hyphae) provide mechanical support for the algae and can absorb water and minerals from its environment. The algae perform photosynthesis and contribute essential sugars for the nutrition and growth of both.

Mya Abbreviation for *million years ago*.

nappe A recumbent (flopped over) fold in rock layers that has sheared through so that its upper limb has been forced far forward.

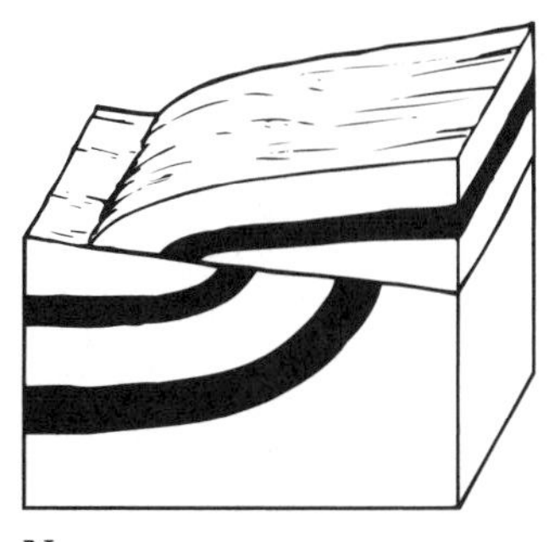

Nappe

native elements Chemical elements that occur in the Earth's crust in a free and uncombined state. They are gold, silver, platinum, copper, lead, tin, zinc, mercury, tantalum, bismuth, antimony, arsenic, selenium, tellurium, sulphur, and carbon (as diamond or graphite).

natural gas An important non-renewable energy source consisting of about 85% methane and up to about 10% of ethane. Smaller proportions of butane, pentane, and other hydrocarbons in the paraffin series may be present.

neap tide The exceptionally low high tide or high low tide that occurs on most coasts twice a month.

nebula A dust and gas cloud in space; a source of stars.

nekton Free-swimming water animals, as opposed to those passively transported by moving water. *See also* zooplankton.

nematodes Members of phylum Nematoda, unsegmented worms, containing many classes.

Neogene period The second part of a revised two-part division of the Cenozoic era. It ran from the Miocene epoch to the Pliocene epoch, about 24–2 million years ago.

Neptune A cold, gaseous planet four times the diameter of the Earth and 30 times as far from the Sun.

neutron A type of subatomic particle with no electric charge. Neutrons and protons form an atom's nucleus and make up almost all its mass.

neutron star The smallest but densest kind of star, apparently resulting from a supernova explosion that left only a compact mass of subatomic neutrons. A neutron star 15 miles (25 km) in diameter can equal the Sun's mass.

night The period of darkness at a place while it lies in the Earth's shadow.

nimbostratus A thick sheet of cloud with a low base that sheds steady rain or snow. Nimbostratus often marks the passing of a depressional warm front.

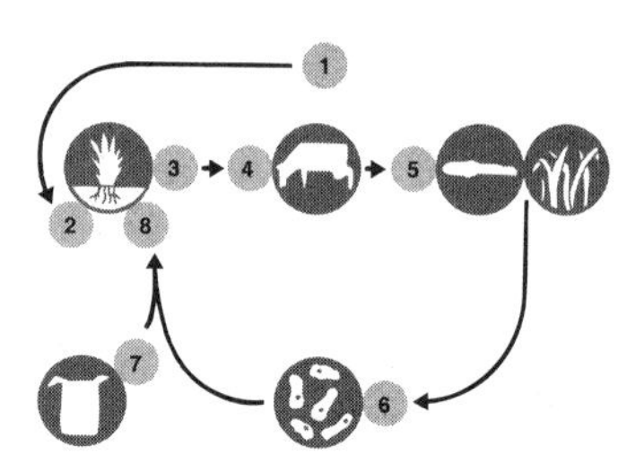

Nitrogen cycle
1 Nitrogen in the air
2 Some plant roots trap nitrogen (nitrogen fixing)
3 Plants use nitrogen to make proteins
4 Animals eat plant proteins
5 Bacteria convert proteins of dead organisms into ammonia
6 Other bacteria convert ammonia to nitrates
7 Artificial nitrates added to soil as fertilizer
8 Plants absorb nitrates

nitrate A nitrogen compound essential for plant growth but liable to contaminate water supplies when nitrate fertilizer is washed into rivers or seeps underground.

nitrogen A chemically inert, colorless and odorless gas making up about 80% of the Earth's atmosphere. Nitrogen is a constituent of animal and other proteins, and the Earth's nitrogen cycle is essential for life.

nitrogen cycle The natural circulation of nitrogen (including nitrogen compounds) from air to consuming organisms (plants, animals, and bacteria) and back to air.

Northern Hemisphere The half of the Earth north of the equator.

North Pole (1) The geographic North Pole, 90° north: an imaginary point in the Arctic Ocean at the northern end of the Earth's axis.
(2) Magnetic north pole: a variable point in the Arctic to which one end of a compass needle points. At the magnetic pole itself the needle tries to point downward.

nova A star that briefly glows intensely brightly as it throws off a little of its substance.

nuclear energy or **atomic energy** The most powerful known form of energy, produced by the fission or fusion of the nuclei of atoms. *See* nuclear fission; nuclear fusion.

nuclear fission The splitting of heavy atomic nuclei to release energy in the form of heat and radioactivity. The nuclear fission of uranium nuclei fuels nuclear power stations and atomic bombs.

nuclear fusion The fusion of lightweight atomic nuclei to form nuclei of a heavier element and to release immense quantities of energy. Thermonuclear reactions, producing helium from hydrogen, fuel stars and are the basis of hydrogen bombs.

nuclear waste Radioactive waste produced by the nuclear industry. Safe, long-term disposal of this waste is a major problem.

nutation Irregularities in the orbital motion of a planet such as Earth other than that caused by the precession of the axis.

obsidian A glassy volcanic rock.

occlusion In an atmospheric low-pressure system, a weather front where a cold front undercuts a warm front from behind, forcing the warm air to rise.

occultation The disappearance of a star or planet behind the Moon, or of any planetary satellite behind the parent body. Occultation of stars by the Moon can be used to determine longitude.

ocean The great mass of saltwater surrounding the Earth's landmasses, or any of its four subdivisions: the Pacific, Atlantic, Indian, and Arctic Oceans.

ocean current A horizontal flow of water through the ocean. Warm and cold surface currents help to redistribute the Sun's heat around the Earth. Cold, dense, polar water feeds deepwater currents that flow back toward the Tropics.

oceanic crust The Earth's crust beneath the ocean basins. Oceanic crust is thinner and denser than continental crust.

oceanic trench A deep, narrow trough in an ocean floor, where oceanic crust is being subducted into the mantle.

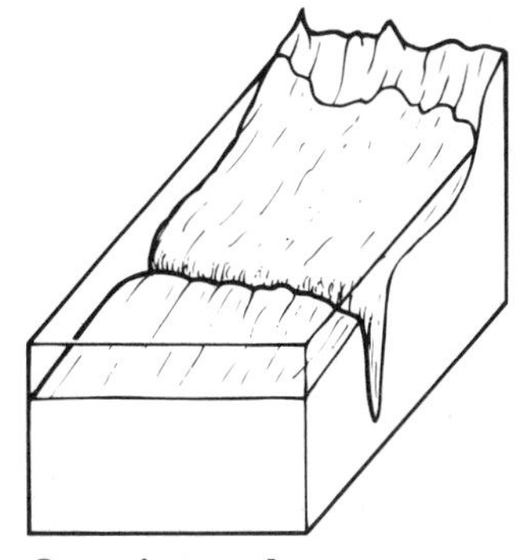

Oceanic trench

oceanography The science of oceans and the organisms living in them. Oceanography includes the study of the chemical composition of the water; the physics of the sea and seafloor; the tectonics of the seafloor; the study of ocean sediments and rocks; the motion of seawater and its responses to internal and external forces; the interaction of the sea and the atmosphere; the living content of the seas and seafloors; the biology of marine organisms; and the formation and interaction of shores, beaches, and estuaries. It also incorporates the marine aspects of a range of other disciplines, including biology, chemistry, physics, geology, meteorology, geophysics, geochemistry, and fluid mechanics.

Octopoda The order of cephalopods that contains the octopus and about 150 other similar species. The Octopoda have eight arms equipped with suckers.

Odonata The order of insects that contains the dragonflies, of which there are some 3,000 species.

oil *See* petroleum.

oil trap The combination of factors that leads to the accumulation of oil, preventing its further movement toward the surface.

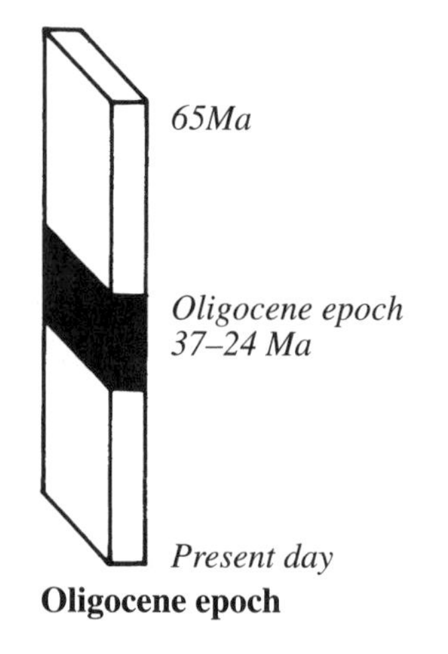

Oligocene epoch

Oligocene epoch The third epoch in the Tertiary period of the Cenozoic era. The Oligocene lasted about 13 million years, from 37 to 24 million years ago (Ma).

olivine A type of silicate mineral found in many igneous and some metamorphic rocks. The transparent green form of olivine, called peridot, is prized as a gemstone.

onyx Banded quartz, of which the variety known as sardonyx, with reddish-brown and white or black straight and parallel bands, is regarded as a gemstone.

oolite Sand consisting of tiny, rounded carbonate grains known as ooids.

oolitic limestone Limestone formed largely from oolite.

opal A hydrated form of quartz (silicon dioxide or silica) valued as a gemstone for its opalescence. Opal may be of many colors and has a hardness of 5–6 on the Mohs' scale.

orbit The curving path of one object revolving around another, for instance, the Earth revolving around the Sun, or the Moon revolving around the Earth.

Ordovician period The second period (505–440 million years ago) of the Paleozoic era. At this time animals still lived only in the sea.

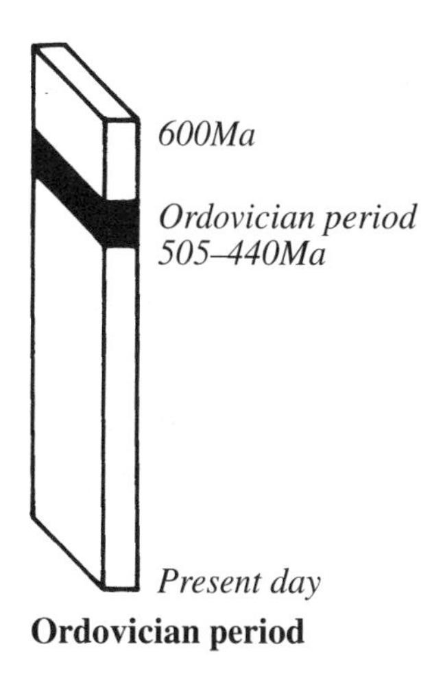

Ordovician period

ore A mineral-rich deposit worth mining. Some ores contain several kinds of metal. Ore deposits are most commonly worked for their metal content, but other minerals, such as fluorite or gypsum, may also be desired. Metal ores occasionally contain native elements, but, most commonly, the metals are present as compounds, especially oxides, sulfides, sulfates, and silicates. Useful ores are diluted with minerals of little value. This waste material is called gangue.

organic Of, or derived from, living things. Describing chemical compounds based on carbon.

organic rocks Sedimentary rocks formed from the remains of living things accumulating under the sea or in swamps. They include coal and limestone.

Ornithischia One of the two orders of dinosaurs. The Ornithischia had bird-like pelves and are believed to have been herbivores, for example, Triceratops. *See* Saurischia.

orogenesis *See* orogeny.

orogenic belt A straight or curved region that has experienced compression tectonics with upthrust. Many orogenic belts are believed to have been involved in subduction of oceanic lithosphere or the collision of major continental masses. They form the sites of the formation of major mountain chains such as the Andes, the Himalayas, and the Appalachian chain.

orogeny A mountain-building phase. During an orogeny, colliding lithospheric plates buckle immense rock masses upward, producing mountain ranges and chains. This process is called orogenesis.

orographic rainfall or **relief rainfall** Rainfall triggered where moist air rising to cross mountains cools and its moisture condenses to form clouds and rain.

orthoclase An alkali feldspar mineral that is an important ingredient in igneous and some metamorphic rocks.

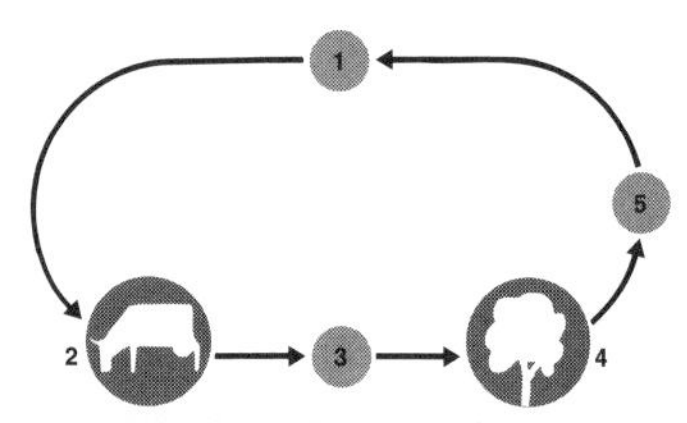

Oxygen cycle
1 Oxygen in the air
2 Oxygen breathed in by animals
3 Oxygen in, carbon dioxide breathed out
4 Carbon dioxide used by plants to make food by photosynthesis
5 Plants release oxygen by day as a result of photosynthesis, if this is faster than the rate of respiration

1985 1987

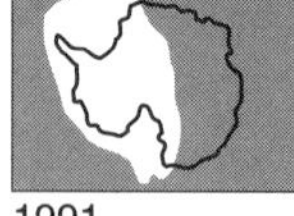
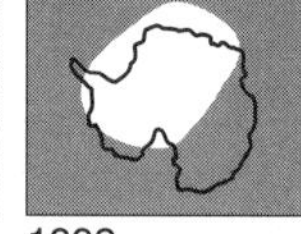

1991 1992

Ozone layer

Orthoptera A large order of insects containing the grasshoppers, crickets, katydids, locusts, stick insects, mantids, cockroaches, and others.

outcrop That part of any body of rock that is exposed at the Earth's surface.

outwash deposits Sediments deposited by streams flowing out of an ice sheet or glacier. Deposits include layered clay, sand, and gravel.

outwash plain A plain with a surface consisting of outwash deposits.

overburden Rock and soil overlying a useful mineral deposit and requiring removal before miners can start strip-mining the deposit.

overfishing Harvesting fish stocks faster than the surviving fish can breed to make good the deficit.

overfold Rock strata folded so that one side of the fold has flopped over above the other side.

overgrazing Letting livestock graze land so heavily that soil loses fertility and grass becomes sparse. Soil erosion may also occur.

oxbow lake A small, curved, narrow lake: a meander cut off from the rest of a river flowing across its floodplain. An oxbow lake forms where the river cuts through the narrow neck of a meander and the ends of the bypassed cutoff silt up.

oxide An inorganic chemical compound consisting of oxygen combined with another element.

oxygen cycle The process by which oxygen circulates between the atmosphere and living things.

ozone A powerfully toxic and unstable gas produced by the action of ultraviolet radiation or electrical discharge on oxygen in air. The molecule consists of a ring of three oxygen atoms (O_3), and only a small quantity is normally present in atmospheric air.

ozone layer A stratospheric layer rich in ozone, a form of oxygen gas that absorbs solar energy and partially shields the Earth from the harmful effects of the Sun's ultraviolet radiation.

pack ice A mass of ice floes (sheets of floating ice) jammed together on the sea. In winter, pack ice covers much of the Arctic Ocean and the Antarctic Ocean.

Paleocene epoch The first epoch (about 65–58 million years ago) in the Tertiary period of the Cenozoic era. Dinosaurs, ammonites, and many marine reptiles were extinct, and mammals began to diversify. Paleocene marine life featured the spread of bivalve mollusks (pelecypods), snails, limpets, slugs, etc. (gastropods), sea urchins (echinoids), and rhizopod protozoa (foraminiferans) that were already present in very small numbers in earlier seas.

paleoclimatology The scientific study of prehistoric climates.

Paleogene period The first part of a revised two-part division of the Cenozoic era: 65–24 million years ago.

paleogeography The scientific study of the Earth's geography in prehistoric times.

paleogeologyThe geology of an area during a specific period of the past.

paleomagnetic dating Dating rocks and fossils by the magnetic alignment of particles in rock. Their alignment is fixed when the rock forms and depends on the Earth's magnetic field, which fluctuates through time.

paleontology The scientific study of fossilized prehistoric plants, animals, and other organisms.

Paleozoic era The first era (about 540–245 million years ago) of the Phanerozoic eon, during which multicellular organisms diversified in the sea and colonized the land. The era is divided into the Cambrian, Ordovician, Silurian, Devonian, Mississippian and Pennsylvanian (Carboniferous), and Permian periods. Paleozoic invertebrates included the brachiopods, trilobites, nautiloid cephalopods, and bryozoans, now either extinct or very rare. The mid-Paleozoic featured the rise of fishes, followed, in the Carboniferous, by amphibians. Vertebrate life in the Permian was dominated by reptiles, some of which were several yards long.

Pangaea breaking up over a period from 135Mya to the present day

Pangaea

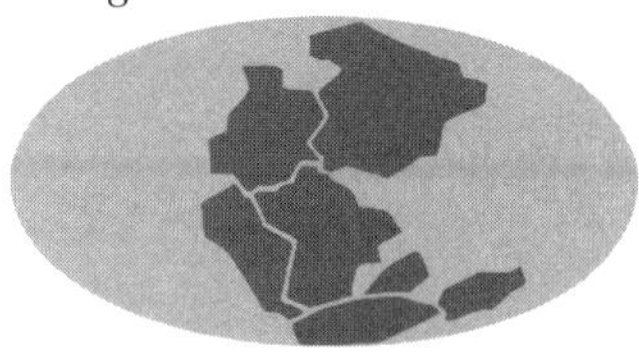

135Mya

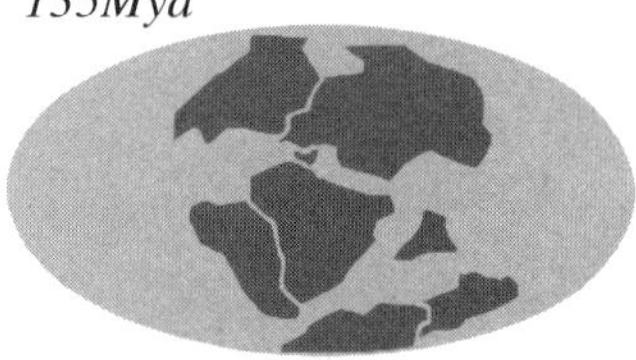

65Mya

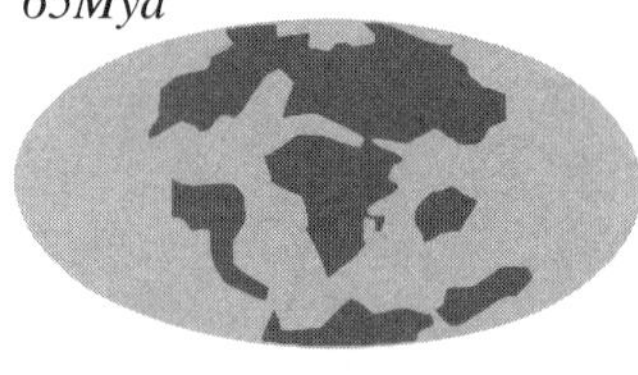

Present day

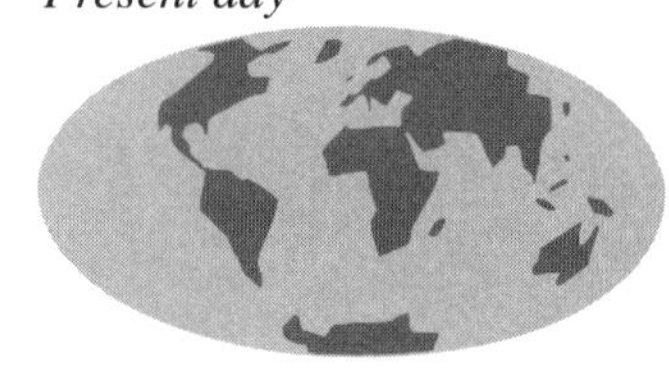

Pangaea A prehistoric supercontinent containing the Earth's major landmasses. It formed late in the Paleozoic era and broke up during the Mesozoic era. *See also* Gondwana; Laurasia.

parallax The apparent difference in position of an object in relation to its background when seen from different positions.

parallel drainage pattern A drainage pattern where a river and its tributaries run largely parallel to each other.

paramagnetism The property of materials, which, when placed in a magnetic field, become magnetized parallel to the field and to a degree that is proportional to the strength of the field. All metals are paramagnetic, as are atoms and molecules with an odd number of electrons, such as free atoms and free radicals.

parasite An organism living on or in the body of another living organism.

parasitism The primary action of a species, from any kingdom, that lives on or in another species (the host), deriving nourishment from it and, in the process, harming it in some way. Every species of living thing is subject to parasitism. Large vertebrates, for instance, are parasitized by lice; these in turn are parasitized by protozoans; protozoans by bacteria; and bacteria by viruses. Ectoparasites live on the outside body surface of the host; endoparasites live inside the host bodies and include worms, flukes, burrowing flies and other insects, fish, fungi, protozoa, bacteria, and viruses.

parsec A unit of astronomical distance equal to 3.26 light-years. It is the distance away a star must be to produce a parallax shift of one second of an arc when viewed from the two extremes of the Earth's orbit around the Sun.

particle size The size of grains in a sediment. There are various classifications, but typically, particles are sized as boulders more than (10 inches – 256 mm), cobble (2.51–10 inches – 64–256 mm), pebble (0.3–2.5 inches – 2–65 mm), sand

(62.5–2000 microns), silt (4–62.5 microns), and clay (less than 4 microns).

patterned ground Polygons, stripes, and other patterns of stones on the surface of periglacial regions. They result from the sorting effects of alternate freezing and thawing.

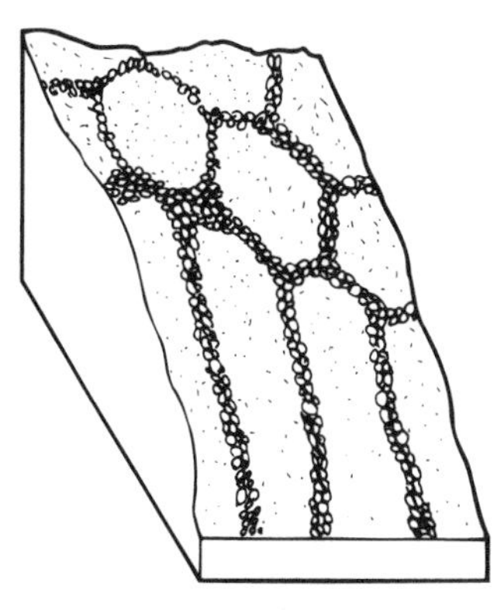
Patterned ground

peak A sharply defined mountain summit.

peat A soft, dark mass of partly decayed plants, such as mosses, sedges, trees, and other plants, that grow in marshes. It occurs largely in moist temperate regions. Dried peat is burned as a fuel and used as a plant-growing medium. When buried and subjected to pressure and heat, forest-type peat is the natural forerunner of most coal.

pebble A rounded stone with a diameter between that of gravel and cobbles.

ped A naturally formed unit of soil structure.

pediment The gentle slope at the foot of a desert mountain. Loose material eroded from the mountain may cover its surface.

pegmatites A class of very coarse-grained igneous rocks, containing much the same mineral ingredients as granites, that form in areas of local concentration of hot gases and fluids in which larger crystals can grow. Quartz and feldspar crystals intergrow in pegmatites. Accessory minerals found in pegmatites include beryl, topaz, fluorite, cassiterite, spodumene, and lepidolite mica.

peneplain A nearly flat land surface almost worn down to sea level by prolonged erosion.

peninsula A tract of land extending out into a sea or a lake.

Pennsylvanian period In North America, the second part of what is generally called the Carboniferous period, about 320–286 million years ago. *See also* Mississipian period.

penumbra (1) An area cast in partial shadow during an eclipse. *See also* umbra.
(2) The relatively pale edge of a sunspot.

peridotite A coarse-grained igneous rock formed mainly of olivine and pyroxene.

periglacial Describing a cold landscape or climate, such as that found in regions near ice sheets. Periglacial features include frost-shattered rocks, patterned ground, and permafrost.

perihelion A planet's closest orbital position to the Sun.

period A geological time unit within an era.

Periodic Table An arrangement of the elements placed in the order of the number of protons in the nucleus of each and divided into rows and columns so as to bring out the similarities in chemical properties.

permafrost The permanently frozen ground of polar and subpolar zones and of any area in which the temperature remains below freezing point for several years. Some 25% of the total land area of the Earth contains permafrost. Most of it lies about 3 feet (1 m) deep, below a so-called active layer that thaws in the brief summer. Permafrost restricts or stops plant growth, prevents most groundwater movement, preserves organic remains, and promotes frost action on rocks. *See also* tundra.

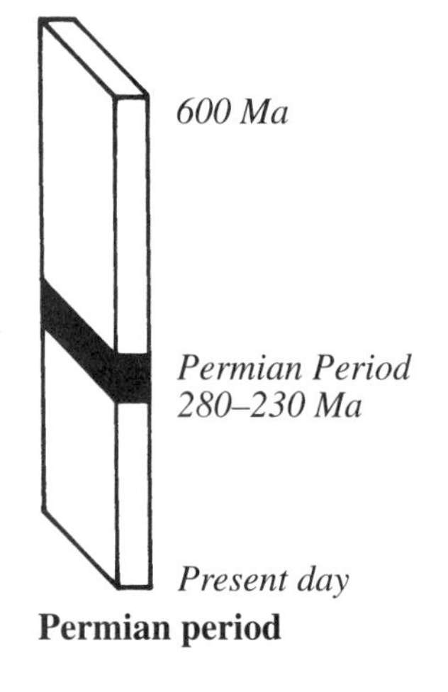

Permian period

permeability The property of a soil, rock, or sediment that allows water or other fluids to pass through a sample of it. Permeability can be quantified in terms of volume flow per unit time under standard conditions of cross section, fluid gradient pressure, and temperature. Permeability is not the same as porosity.

permeable rock Rock that allows water to pass down through cracks or pores in the rock. *See* pervious rock; porous rock.

Permian period The last period of the Paleozoic era, about 280–230 million years ago. During Permian times landmasses formed a supercontinent known as Pangaea. It ended with the greatest known mass extinction of species.

perturbation The deviation or digression of a celestial body from the path it would take if subject only to a single influencing force.

pervious rock Rock with cracks allowing water to pass down through it. *See* permeable rock; porous rock.

pesticides Chemicals that kill plant, animal, and other pests but may also pollute food and water supplies. Well-known toxic pesticides include Aldrin, Chlordane, and DDT.

petrifaction The turning to stone of an organic body. Petrifaction is the process by which fossils are preserved. The organic material becomes impregnated with mineral salts that are deposited from solution.

petrography The formal description of rocks and their identification and classification and the resulting interpretation of their origins. Using a simple hand lens, a skilled petrographer can identify most rocks whose grains are larger than 0.04 of an inch (1 mm).

petroleum (mineral oil) Mineral oil and natural gas (hydrocarbons) formed in the Earth's crust as pressure and heat acted on the remains of billions of marine plants and animals compressed by rocks formed from layers of sediment.

petrology The scientific study of rocks, their physical and chemical properties, and their modes of origin. Petrology is concerned with the three classes of rocks – sedimentary rocks, made from settled fragments of preexisting rocks, organic products, or chemical precipitates; igneous rocks, which consist of solidified molten matter (magma) from below the Earth's crust; and metamorphic rocks, derived from igneous or sedimentary rocks, or both, that have been changed in their mineral composition, texture, and internal structure by heat, pressure and other influences.

Phanerozoic eon The "age of visible life," the fossil-rich past 540 million years of Earth's history.

phases of the Moon The apparent change in shape of the Moon from new moon to full moon and back, every 27.31 days. The amount of the Moon's sunlit side visible from Earth varies as the Moon orbits the Earth.

Phases of the Moon

phenocryst A large, readily visible, and usually well-formed crystal surrounded by smaller crystals in igneous rock.

photon A particle of light. When lightwaves hit matter, they behave as energy particles rather than waves.

photosphere The visible surface of the Sun or another star. Most of the sunlight reaching the Earth comes from the photosphere.

phyllite A fine-grained metamorphic rock formed from sedimentary rocks, such as mudstone and shale. It tends to split into sheets.

phylogeny The sequence of events involved in the evolution of a species. The history and lineage of organisms and the timescale of their evolution.

phylum (pl. **phyla**) A major taxonomic division of the animals and plants that contains one or more classes. A phylum is below a kingdom and above a class. The taxonomic divisions are kingdom, phylum, class, order, family, genus, and species.

physical geography Study of the Earth's features to gain an understanding of our environment and interactions between people and the environment. Special studies include geology, geomorphology, climatology, and oceanography.

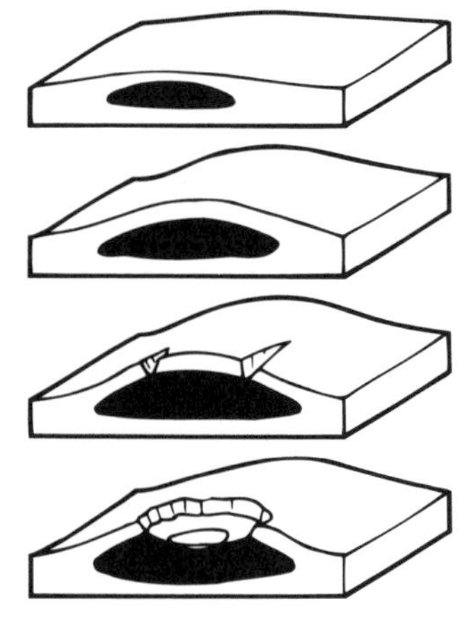
Pingo

physical weathering *See* mechanical weathering.

phytoplankton Drifting plants and plantlike organisms, especially those that teem in the surface waters of oceans. Almost all sea creatures directly or indirectly depend on phytoplankton for food. *See also* plankton; zooplankton.

pie chart A circle divided into segments representing parts of a whole; for example, a world population pie chart with segments of varying sizes representing the populations of individual continents.

pingo A hillock produced in periglacial regions by an underground ice blister pushing up the surface materials.

placer deposit A deposit of minerals that has been concentrated by natural processes of weathering, as in the transport and deposition of gold by a stream.

plagioclase A group of plentiful feldspar minerals forming pale, glassy crystals – important ingredients in certain igneous and metamorphic rocks.

plain A large tract of almost level lowland. Examples include North

America's Great Plains, Eurasia's steppes, and South America's pampas.

planetesimals The "mini planets" that are believed to have collided and coalesced to build planets such as the Earth.

planets Large objects orbiting the Sun or another star. Our solar system contains nine planets.

plankton Mostly small to minute organisms passively drifting around in immense numbers in oceans, lakes, etc. *See also* phytoplankton; zooplankton.

plateau A large area of high land with a fairly flat top. Examples include Asia's Tibetan Plateau and South America's Andean plateau region, known as the *altiplano*.

plate boundary or **plate margin** A boundary between two lithospheric plates. At constructive plate boundaries, spreading ridges produce new oceanic crust. At destructive plate boundaries, crust is subducted below oceanic trenches.

plate margin *See* plate boundary.

plate tectonics The now widely accepted theory that the Earth's crust consists of moving lithospheric plates lying above a weaker semiplastic asthenosphere and that their interactions build and destroy continents and oceans. Plate tectonics theory provides a convincing explanation for the present-day tectonic behavior of the Earth, its continental drift, the distribution of mountain ranges, seafloor spreading, earthquake activity, the occurrence of volcanism in a series of linear belts, and the observed anomalies of the magnetic patterns of the seafloor. Adjacent plates may move apart from one another, or toward one another, or may slip past one another. The speed of plate movement relative to each other ranges from 0.8–8 inches (2–20 cm) per year.

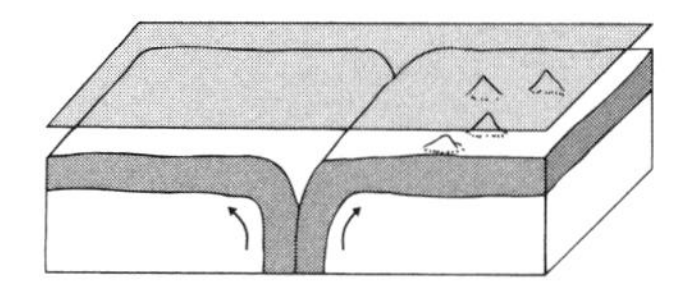

SPREADING
Underwater volcanoes on ocean floor

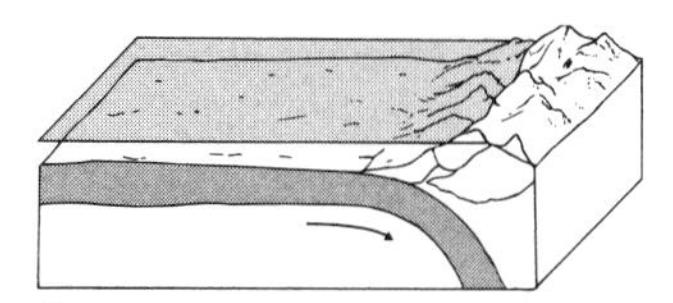

SUBDUCTION
Plates push against each other and one bends downward

Plate tectonics

Platyhelminthes The phylum of invertebrates known as the flatworms. They include many flukes and tapeworms that are parasitic on humans.

playa A desert depression sometimes filled by a salty lake that later evaporates.

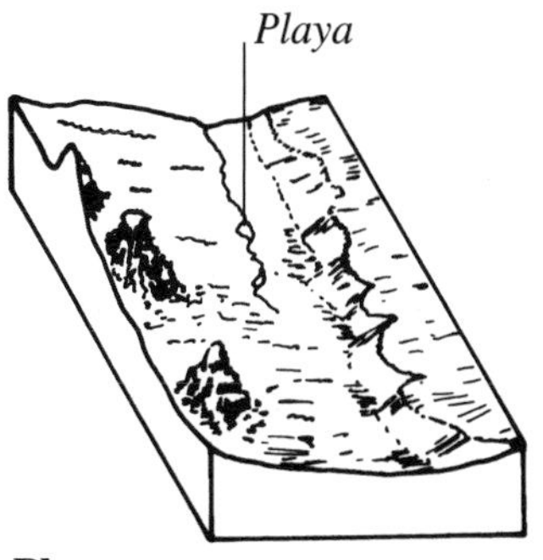

Playa

Pleistocene epoch The first part of the Quaternary period, about 2 million to 10,000 years ago. For much of this time ice sheets covered large parts of northern continents, but the Pleistocene featured at least 17 worldwide cyclic changes of climate. Landforms in many parts of the world date from this period.

Pliocene epoch The last part of the Tertiary period, about 5–2 million years ago.

Pluto The smallest, coldest planet of the solar system, usually the farthest from the Sun.

pluton A general term for any large mass of igneous rock, such as granite formed from magma that cooled deep in the Earth's crust.

poikiloblast A large crystal that encloses smaller crystals in metamorphic rock.

polar Of or relating to a planet's poles.

polar front The shifting boundary between polar and tropical air masses where mid-latitude and high-latitude low-pressure systems originate.

polar regions Regions around the North Pole and South Pole.

polar wandering The prehistoric wandering of the Earth's magnetic poles, revealed by the different magnetic alignment of particles in rocks formed at different times.

polarity or **geomagnetic polarity** The magnetic alignment of certain particles in igneous rocks, determined by the Earth's magnetic field at the time the rocks formed.

Poles The ends of the Earth's axis, forming its northernmost and southernmost points: the North and South Poles. Their locations do not correspond exactly with those of the (variable) north and south magnetic poles produced by the Earth's magnetic field.

pollutant Anything that pollutes.

pollution Harmful substances introduced into the environment, especially poisonous substances introduced into water and air.

population A group of people, animals, or other living organisms whose numbers are liable to change through death, birth, and migration.

porosity A measure of the total of spaces between the grains in a rock, expressed as a percentage of the total volume of the rock. Permeability is affected by the degree of the interconnection between these spaces.

porous rock Open-textured rock that lets water pass down through tiny pores in the rock. Examples include sandstones and some limestones. *See* permeable rock; pervious rock.

porphyroblast A large, well-formed crystal produced by recrystallization and surrounded by a finer-grained matrix of much smaller crystals in a metamorphic rock.

porphyry Medium to coarse-grained intrusive felsic igneous rock containing more than 25% of large and well-formed crystals (phenocrysts) set in a finer-grained matrix.

potassium An element of the alkali metal group. Found in igneous rocks, it is the seventh most plentiful element on Earth.

pothole A round hole worn in solid rock by water whirling stones around the bed of a fast-flowing stream.

Precambrian All Earth's history predating the Cambrian period: the Hadean, Archean, and Proterozoic eons. The Cambrian is the oldest geologic period from which abundant readily visible fossils have been recovered. In the Precambrian, only microscopic fossils and those of blue-green algae are found. There is evidence that the Precambrian lasted for about 3.5 billion years and ended about 600 million years ago. The rocks of the period are of great variety but differ little from one continent to another. They contain large quantities of metal ores, especially iron, nickel, gold, uranium, and copper.

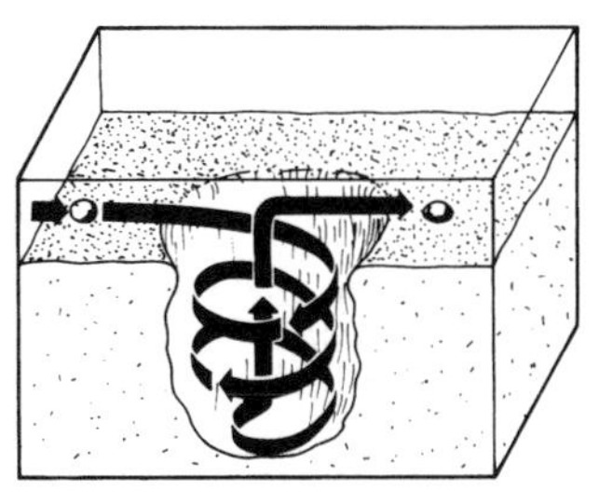

Pothole

precession In astronomy, an apparent slow shift of constellations' positions produced by the Earth's slow-motion wobble, which causes a gradual change in the direction of the Earth's axis.

precious stones Minerals and other substances valued for beauty, rarity, permanence. They include amber, beryl (emerald, aquamarine), corundum (ruby and sapphire), diamond, feldspar (moonstone), garnet, jade, jet, lapis lazuli, malachite, opal, amethyst, citrine, agate, topaz, tourmaline, turquoise, and zircon.

precipitate An insoluble solid that has separated from a solution.

precipitation Water deposited from the atmosphere in the form of rain, hail, sleet, snow, dew, and frost. These are all grouped under the term *hydrometeors*. The amount of precipitation, often simply called rainfall, is measured in a collection gauge and is the actual depth of liquid water that has reached the ground, and, for frozen forms, after they have been melted. It is recorded in inches and hundredths of an inch or in millimeters falling in a given period. The area of the gauge mouth is irrelevant but must be the same as the cross section of the measuring jar.

predation The process of feeding by animals on other species of animals. The idea of the ecological pyramid necessarily involves predation, in which the predator, hunts, kills, and eats the other, the prey. Predation is experienced by most animal species, but, since many animals are herbivorous, the proportion that are actual predators is less. The more successful the predator species, the higher up the ecological pyramid it will be situated. Top predators are those at the apex of the food chain pyramid (such as humans) and are not, themselves, generally preyed upon.

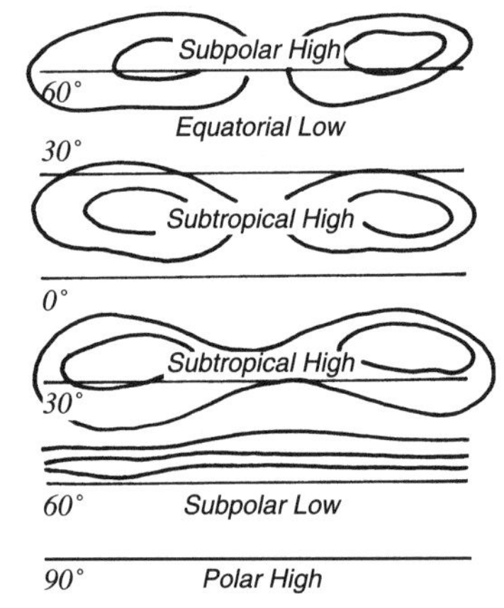

Pressure belts

pressure belts Global zones of high and low atmospheric pressure. They depend largely on regional temperature differences and to some extent shift north and south with the seasons.

pressure dissolution The process that occurs at the contact interfaces of crystals or grains as a result of pressure, bringing about the dissolving of material and compaction.

pressure system A large rotating mass of high or low pressure air. *See also* anticyclone; cyclone.

primary wave *See* P wave.

prime meridian An imaginary line of longitude that passes through

Greenwich, England – a reference line for the longitudes of all places on Earth, which are measured in degrees east or west of the prime meridian.

profile (long profile) The slope of a river from source to mouth. When deposition balances erosion, the result is a smooth long profile known as a graded profile.

prominence (solar) An explosive eruption of gases from the Sun's surface.

prospecting Searching parts of the Earth's crust for useful mineral deposits or for fossils.

Proterozoic eon The third eon (2.5 billion–540 million years ago) in Earth's history. Large continents formed and multicellular organisms appeared in the sea.

proton A positively charged type of subatomic particle. Protons and neutrons form an atom's nucleus and make up almost all its mass.

protostar In space, a mass of dust and gas condensing by gravitational attraction in the course of forming a star.

pumice A glassy rock formed when the froth or crust on the surface of very gassy acidic lava hardens. Pumice is thus honeycombed with holes of various size and, for a rock, is of unusually low density.

P wave or **primary wave** A fast-moving compressional earthquake wave, the first to reach a seismic observatory. They travel about 1.7 times faster than S waves. *See also* S wave; L wave.

pyroclastic Of volcanic rocks that consist of fragmented particles usually produced by the natural explosive action of expanding volcanic gases.

pyroxene Any of a group of dark, dense, rock-forming chain silicate minerals rich in calcium, iron, and magnesium. The group includes augite, hypersthene and diopside.

quartz A hard mineral made of silica (silicon dioxide), a major

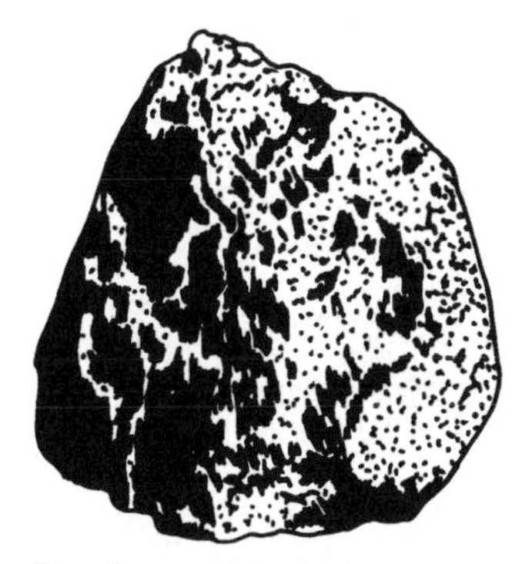
Pumice

constituent of most sand, sandstone, and of many other rocks. Quartz is the second most abundant mineral after feldspar. Its hardness is designated as 7 on Mohs' scale, and its specific gravity is 2.650. It is a constituent of almost all types of igneous, sedimentary, and metamorphic rocks, and occurs in meteorites. Quartz is the principal mineral of many gemstones, including amethyst, citrine, smoky and rose quartz, jasper, and agate. Quartz crystals exhibit piezoelectricity, and are used in millions of electronic circuits, such as those of computers, requiring stable oscillators.

quartzite A metamorphic rock made of quartz derived from sandstone.

quasars Short for quasi-stellar sources. These are very remote but extremely bright objects in space, thought to mark the early formation of galaxies.

Quaternary period The second (present) period of the Cenozoic era. It began about 2 million years ago and encompasses the Pleistocene and Holocene epochs. By the Pleistocene, most flora and fauna were of modern pattern. The dominant features of the Quaternary were the ice ages of the Pleistocene in the Northern Hemisphere.

radial drainage pattern A pattern created where streams radiate in all directions from a mountain or upland region.

radiation zone (solar) The region deep in the Sun where radiation produced by nuclear reactions in its core moves freely outward.

radioactivity The emission of rays and subatomic particles from the nuclei of certain elements and their decay products, notably uranium and its decay products, down to but excluding lead. Radioactivity from nuclear bombs and installations and even certain rocks can injure living tissues.

radioactivity exploration The location of radioactive elements, such as uranium and thorium, using geiger counters and scintillation counters as detectors. Both land and airborne surveys are performed. In the latter case, because natural radiation from buried rocks is usually absorbed by a few feet of soil cover, general measurements of diffuse equilibrium radiation are made.

radio astronomy The collection and analysis of radio waves reaching Earth from space.

radiocarbon dating A method of estimating the age of organic materials based on the decay of the isotope carbon-14. This isotope decays with a half-life of approximately 5,700 years and is produced by cosmic-ray bombardment of the upper atmosphere. Living things turn over carbon-14 so that, long-term, it is present in roughly constant levels in most of the biosphere. But once an organism dies, carbon-14 is no longer assimilated, and the level of this isotope in it begins to decrease at the rate determined by its half-life. So a measurement of the residual carbon-14 activity in a sample compared to the general level enables the age to be determined. The method is used to date organic materials less than 70,000 years old.

radiometric dating Dating rocks by the known rate of decay of radioactive elements that they contain.

radio waves Electromagnetic waves moving through air and space at the speed of light and carrying signals that can be converted to sounds.

rain Falling water drops formed from droplets coalescing in clouds.

rainbow A colored arch seen in the sky when sunlight shines through raindrops, splitting the light into bands of its component colors.

rainfall The total precipitation (rain, dew, melted snow, frost, hail, and sleet) measured by a rain gauge over a certain specified period.

rain gauge An instrument measuring rainfall. A measuring cylinder shows the amount of rain that has run down a funnel into a collecting container.

rainwash or **sheet erosion** Erosion of the ground surface by rainwater flowing down a smooth slope.

raised beach A beach that stands above today's sea level because the sea level has fallen or the land has risen.

raised coastline A coastline with raised beaches or other features showing that the land has risen or the sea level has dropped .

rapids A stretch of turbulent water flowing quickly over a stream bed that suddenly steepens, as when it crosses a band of resistant rock.

rare earth Any metallic element with an atomic number between and including 58 and 71. Rare earths, also called lanthanides, are chemically similar to actinides (89 through 104).

ravine A narrow, steep-sided valley smaller than a canyon but bigger than a gully. Ravines form in dry lands subject to occasional heavy rain.

Rayleigh wave A type of long wave (L wave) triggered by earthquakes and moving through the Earth's surface. It is named for the English physicist and mathematician Lord Rayleigh.

recessional moraine A ridge of rock debris dumped by the melting front of a glacier or ice sheet as it retreats.

recrystallization The growth of new mineral grains from earlier ones by the diffusion of ions under the influence of pressure and temperature.

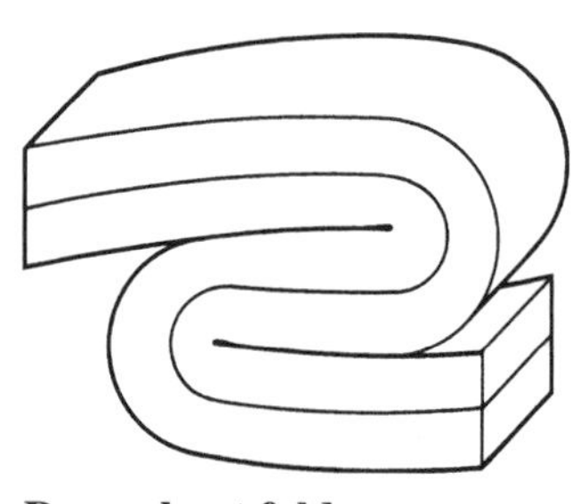

Recumbent fold

recumbent fold Layered rocks crumpled into a fold with one side flopped over on top of the other.

recycling Conserving by reusing (directly or after reprocessing) articles made from substances such as aluminum, glass, and paper.

red giant A star that is 10 to 100 times the size of the Sun.

red shift The shift toward the red end of the spectrum in light reaching the Earth from remote stars moving away from it. The red shift is a Doppler phenomenon most simply understood as the "stretching" to longer wavelengths of waves emitted by a source moving away from the observer. The red shift, noted by Edwin Powell Hubble, is the basis for the belief in the expanding universe.

reef A ridge of rock or coral always or often just submerged by the sea.

regional metamorphism Recrystallization of preexisting rock over the broad area in which lithospheric plates are converging so as to produce a wide area of temperature and pressure changes.

regolith A layer of loose material (soil, subsoil, and broken rock) covering bedrock.

rejuvenation Renewed erosional attack by a river on land after it has been uplifted or after a fall in sea level. Deepening of the river's bed begins at its mouth and works its way back upstream. *See also* knickpoint; terrace.

relative age The placement in time of a geological event as occurring prior to or subsequent to other events.

relief Differences in height for any area of the Earth's surface.

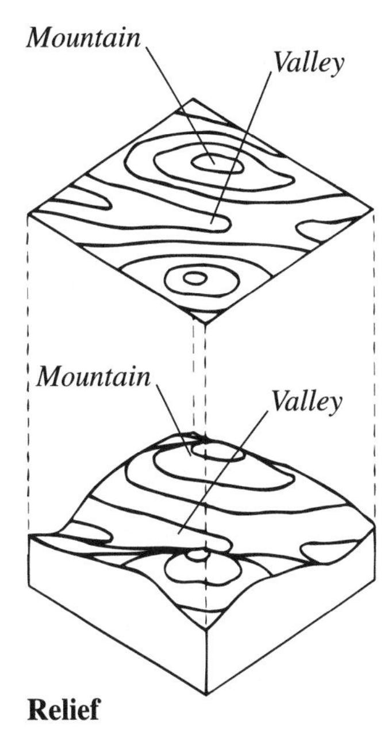

Relief

relief maps Maps that show hills and valleys by means of contours, hachures, layered coloring, etc. *See* contour maps; hachures.

relief rainfall *See* orographic rainfall.

remote sensing Detecting or measuring an object without touching it, especially using satellites or aircraft to map the Earth's surface features.

renewable resources Crops, fish, timber, sunshine, wind, and other sources of food or energy that can be used without exhausting them.

resistant rock Rock that resists weathering and erosion because of factors such as its hardness and insolubility in water.

respiration A process by which living things take oxygen from their surroundings, use it in chemical reactions that produce energy, and release it combined with carbon as carbon dioxide gas.

resurgent stream A stream that reappears after flowing underground through a cave. Resurgence is a feature of some mountainous limestone regions. *See also* karst.

rhyolite A range of very fine-grained extrusive acid igneous rocks, having the same mineral content as intrusive granites. Rhyolites are very hard and break like glass. Some show flow banding or may contain vesicles or amygdales. Accessory mineral content, often as phenocrysts, include garnet, topaz, zircon, pyrite, fluorite, and apatite. Glassy rocks associated with rhyolites include obsidian, pitchstone, and pumice.

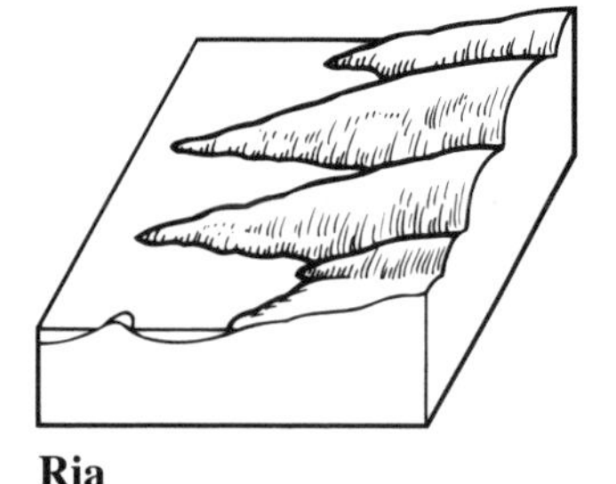

Ria

ria A deep, funnel-shaped inlet of the sea occupying a drowned river valley in a submerged upland coast.

Richter scale A scale measuring the energy released by an earthquake and invented by American seismologist Charles Richter in 1935. On its scale each number represents ten times the energy of the number below it.

rift valley A long, steep-sided, flat-bottomed depression in the Earth's surface. Rift valleys form where a slab of continental or oceanic crust slips down between two or more parallel faults.

rille A long, narrow furrow in the Moon's surface, once occupied by flowing lava. A large rille can be 1,000 feet (300 m) deep and 1 mile (1.5 km) wide.

river A large, natural freshwater stream flowing downhill through a long channel to another river or to a lake or the sea. A river and its tributaries form a river system.

river capture One river capturing part of another. By headward erosion, the first river invades a neighboring valley, intercepts the other river, and captures its water above the point where they meet.

river terrace *See* terrace.

roche moutonnée A rock mound that a passing glacier has shaped like a French lawyer's lambswool wig. Stones in the glacier give it a smooth, gently sloping upstream side. The plucking effect of freezing meltwater gives it a ragged, steep downstream side.

rock Any mass of mineral particles forming part of the Earth's crust. Solid rock includes granite and limestone. Unconsolidated rock includes clay and sand. The three major types of rock are igneous, metamorphic, and sedimentary. *See* igneous rock; metamorphic rock; sedimentary rock.

rock age determination Methods of measuring the age of rocks. The principal method is based on the fact that many rocks and minerals contain radioactive elements that decay spontaneously to form other stable elements. The latter can, under certain conditions, accumulate within mineral crystals so that the proportion of stable elements to radioactive elements increases with time. This ratio can be accurately measured with a mass spectrometer, and, because the half-lives of the parent elements are known, the age of the rock can be calculated. Other methods of dating include stratigraphy, index fossil methods, rock magnetism, geochronometry, fission track dating, dendrochronology, and tree-ring hydrology.

rock cycle The processes by which magma becomes igneous, sedimentary, and metamorphic rock, and these become magma.

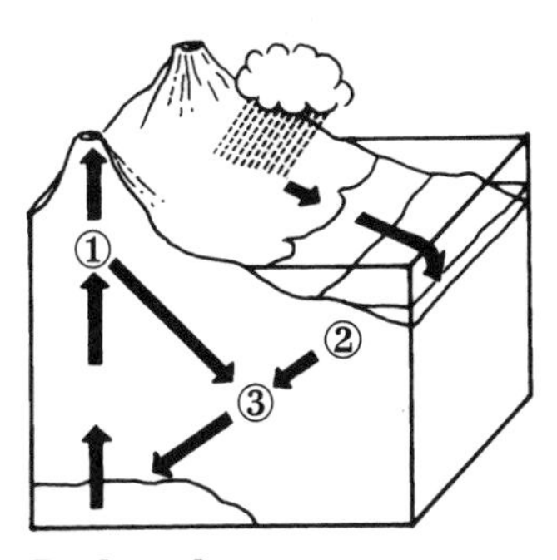

Rock cycle

rockets Engines moved at high speed by the thrust produced by burning the fuel and oxygen that they carry.

rockfall The free fall of rocks from a cliff face. *See also* landslide; rockslide.

rock pedestal A mushroom-shaped rock, its base whittled away by the sandblasting effect of windborne sand. Rock pedestals are a feature of some deserts.

rockslide A mass of loose, weathered rock sliding down a line of weakness in the underyling rock of a slope.

rock step A short, steep step in the floor of a glaciated valley. Ice falls on glaciers suggest underlying rock steps.

rock-stratigraphic units Sedimentary and volcanic rocks identified by their characteristics. The units include (from major to minor): groups, formations, and members.

Rossby waves Large, high-level undulations in the jet streams of the Earth's atmosphere. They separate warm, tropical air from cold, polar air. They are named for the Swedish-American meteorologist C. Rossby.

rotational slippage A landslide where a mass of rock tilts back as it slides down a slope, so that its base moves farther forward than the rest.

ruby Aluminum oxide (corundum) in its rare, deep red-colored form.

rudaceous rock *See* conglomerate.

runoff (1) Rainwater or meltwater flowing through streams and underground to reach the sea.
(2) Rainwater or meltwater running over the surface of the land.

salinization Accumulation in the upper soil levels of salts sucked up by the evaporation of soil moisture. Salinization is a problem in some irrigated soils of hot, dry climates. The soil becomes too salty to grow crops.

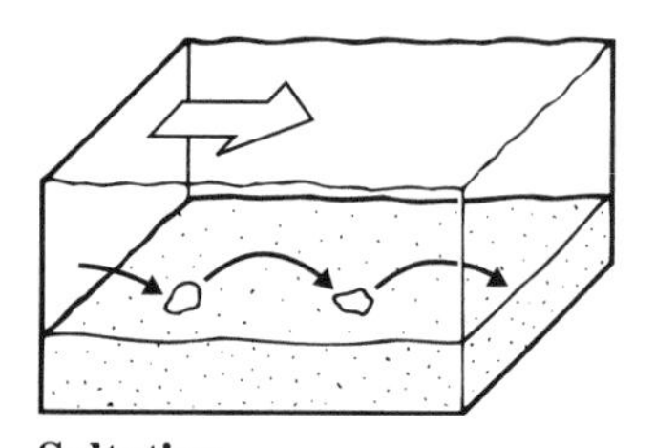
Saltation

saltation The hopping of windblown sand grains over the ground or of tiny waterborne stones along the bed of a stream.

salt dome A massive, vertical underground cylinder of rock salt risen through sedimentary rocks and forcing the rock layers above it up into a dome. Salt domes originate as evaporites.

salts In geology, the class of chemical compounds that are precipitated when seawater evaporates.

sand Rock particles with sizes between those of gravel and silt. Most sand consists of quartz.

sandstone A sedimentary rock chiefly formed of naturally cemented sand grains. There are three major groups: terrigenous, carbonate, and pyroclastic. Terrigenous sandstones are made of clasts derived from rock erosion outside the area of sedimentation and transported by wind and water. Carbonate sandstones come from skeletal debris and locally derived carbonate rock debris. Pyroclastic sandstones are composed of rock fragments produced by explosive volcanic activity. Ten to 20% of the volume of sedimentary rock in the earth's crust is sandstone.

sapphire The mineral corundum in sufficiently attractive form to be regarded as a gemstone. Sapphire may have a wide range of colors and has a hardness of 9 on Mohs' scale.

satellites Natural and artificial moons revolving around planets. Earth satellites include meteorological satellites providing valuable data on world weather conditions.

saturation (1) The condition of air containing all the water vapor it can hold. The amount varies with temperature and pressure.
(2) The condition of rocks containing all the water that they can hold.

Saturn The second largest planet in the solar system, with a diameter almost 10 times that of the Earth. More than 20 moons and 7 rings comprising thousands of pieces of ice orbit this gaseous planet.

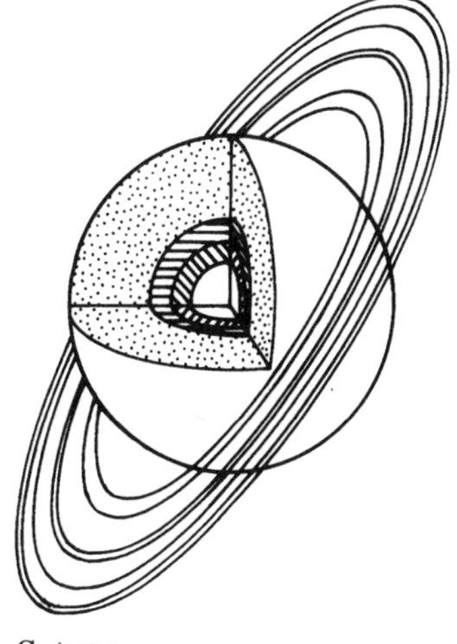

Saturn

Saurischia One of the two orders of dinosaurs. They included bipedal carnivores (such as Tyrannosaurus) and quadripedal herbivores such as Brontosaurus, now named Apatosaurus.

scarp *See* escarpment.

schist A large group of coarse-grained metamorphic rock rich in flaky minerals aligned in parallel bands. Schist can be formed from slate or basalt. As the name implies, schists have zones of more or less perfect cleavage along which they can easily be split up into flat slabs. *See also* schistosity.

schistosity A characteristic of certain types of metamorphic rocks, notably schists and phyllites, to be able to be split along parallel planes. This property comes from the distribution and parallel arrangement of platy mineral crystals such as biotite, chlorite, muscovite, talc, and graphite. Schistosity is also conferred by the presence of rodlike crystals such as actinolite, hornblende, and tremolite.

Schuler pendulum Anything that swings under the influence of gravity with a period of 84.4 minutes. This is the period of a pendulum equal in length to the radius of the Earth. Such a pendulum will remain vertical however the pivot may move.

scintillation counter A detector of radiation in which the receipt of a quantum of radiation is signaled by a flash of light that is detected, amplified, and counted.

Scorpionida The order of Arachnida that have a sting at the tip of the tail. There are about 1,000 species of scorpions but only a minority are capable of a dangerous sting.

scree *See* talus.

sea (1) A subdivision of an ocean.
(2) The name given to some large salt lakes. The Caspian Sea is the largest of these.

sea cave A cave worn in a sea cliff by wave action widening cracks or other weaknesses in the rock.

sea cliff A cliff where land meets sea. A sea cliff starts as a wave-cut notch undermining the foot of a slope. Collapse of the slope's undermined rocks results in a cliff that gradually retreats inland under attack from the sea.

seafloor spreading The theory that magma welling up at submarine spreading ridges creates new ocean crust and this moves away from the ridges at about 1–2 inches (2.5–5.0 cm) a year.

sea level (mean sea level) The average level of the sea surface.

seamount A submarine volcanic peak. *See also* guyot.

seasons Times of year with distinctive weather determined by changes in the angle at which the Sun's rays reach the Earth. Many tropical regions have only two seasons: wet and dry. Temperate regions have four seasons: spring, summer, autumn (fall), and winter. The Northern Hemisphere's spring and summer coincide with the Southern Hemisphere's autumn and winter.

secondary waves *See* S waves.

sediment Any fine or coarse material deposited by water, ice, or wind. Sediment includes dust, silt, sand, gravel, pebbles, cobbles, boulders, and organic remains.

sedimentary cycle The sequence of rock weathering, erosion, transport, deposition, and burial leading to the formation of a new generation of sedimentary rock.

sedimentary rock Rock formed at or near the surface of the Earth from compacted sediments, often deposited in layers (strata) and hardened by natural cements. Layering, or stratification, is the most important single characteristic of sedimentary rocks, which, although only a small proportion of the whole crust of the Earth, represent 75% of the exposed rocks at the surface. Sedimentary rocks are important sources of oil and natural gas, coal, iron ores, and limestone. They include shale, sandstone, limestone, coal, clay, and rock salt. *See also* clastic rocks; evaporite; organic rocks; sedimentation.

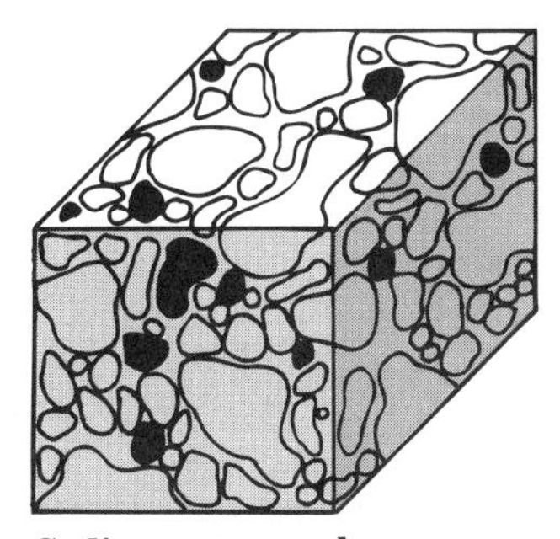

Sedimentary rock

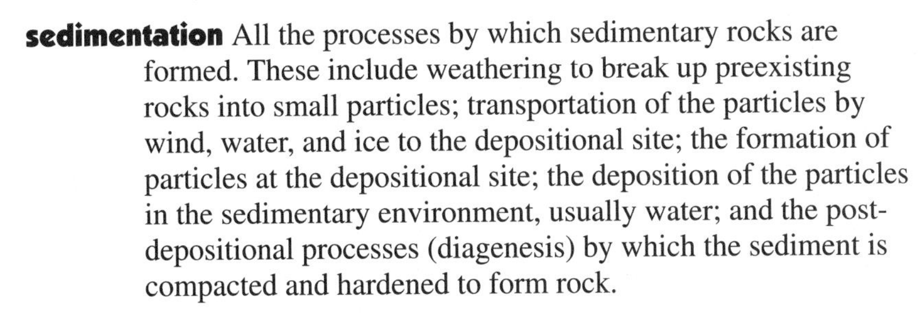

sedimentation All the processes by which sedimentary rocks are formed. These include weathering to break up preexisting rocks into small particles; transportation of the particles by wind, water, and ice to the depositional site; the formation of particles at the depositional site; the deposition of the particles in the sedimentary environment, usually water; and the post-depositional processes (diagenesis) by which the sediment is compacted and hardened to form rock.

seif dune A long sand ridge with a sharp crest, shaped by and aligned with the prevailing wind. Seif dunes occur in deserts. Some contain parallel rows of seif dunes nearly 700 feet (about 210 m) high.

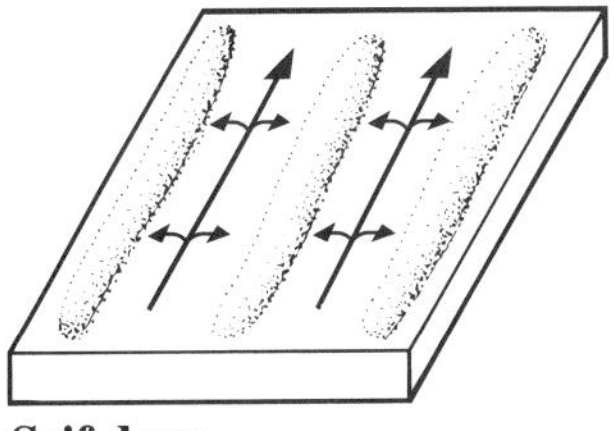

Seif dune

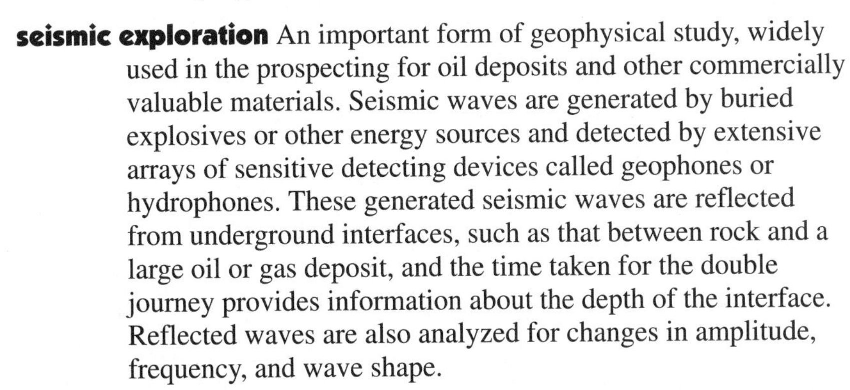

seismic exploration An important form of geophysical study, widely used in the prospecting for oil deposits and other commercially valuable materials. Seismic waves are generated by buried explosives or other energy sources and detected by extensive arrays of sensitive detecting devices called geophones or hydrophones. These generated seismic waves are reflected from underground interfaces, such as that between rock and a large oil or gas deposit, and the time taken for the double journey provides information about the depth of the interface. Reflected waves are also analyzed for changes in amplitude, frequency, and wave shape.

seismic wave Any of several types of waves that ripple out through the ground from the focal point of an earthquake. *See also* tsunami.

seismograph An instrument showing Earth tremors by the difference in movement between a frame that moves with the ground and a suspended weight that stays still. Additional devices amplify the difference, and a pen fixed to the weight records the tremors on paper fixed to a revolving drum. Seismologists record earthquake intensities and help scientists find oil deposits and study the Earth's internal structure.

seismology The scientific study of Earth tremors.

semiarid Describing places located between deserts and grasslands. They feature grasses, shrubs, and small trees adapted to low and irregular rainfall.

series Rocks formed during a geological epoch.

serpentine *See* chrysotile.

shaft mining Mining veins or seams by sinking vertical shafts in the ground from which horizontal tunnels, called levels, branch off to reach the underground coal or mineral deposits.

shale A fine-grained sedimentary rock formed from layers of clay and silt. Shale splits readily into thin, flaky layers.

sheet erosion *See* rainwash.

shelter belt A belt of trees planted to protect cropland from strong winds that would damage crops or blow away the soil they depend on.

shield A geologically stable region of ancient (Precambrian) rocks, often forming a continent's core. Most shields are worn down by prolonged erosion.

shingle Pebbles and other rounded rock fragments forming a beach.

shooting star *See* meteor.

sidereal time Time measured by the Earth's rotation in relation to distant stars.

silica Silicon dioxide. A silicon-oxygen compound, also known as quartz, that comprises 59% of the Earth's crust. It is one of the most plentiful of all minerals. *See also* silicates.

silicates The most plentiful group of rock-forming minerals, usually consisting of silicon and oxygen combined with a metal. Silicates include feldspars, micas, and quartz.

silicon The second most plentiful element in the Earth's crust, found in almost all rocks but always combined with oxygen or with oxygen and other elements including aluminum. The chemistry of silicon is the inorganic analog of the chemistry of carbon.

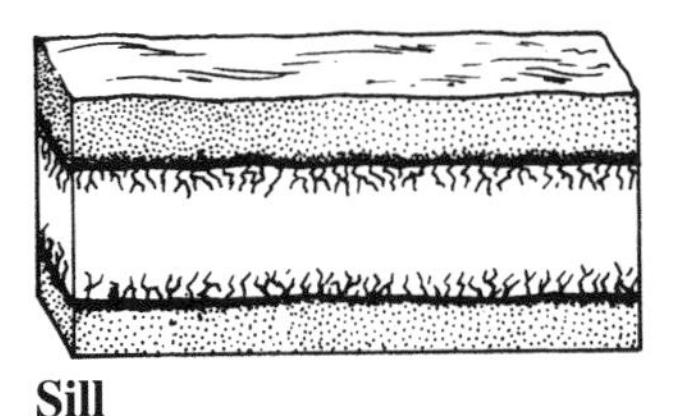

Sill

sill A horizontal sheet of igneous rock intruded between sedimentary rock layers. Some sills extend for hundreds of square miles.

silt Sediment consisting of particles that are bigger than those forming clay but smaller than sand grains. *See also* argillaceous rocks.

Silurian period The third period (about 438–408 million years ago) of the Paleozoic era. Silurian deposits, both terrestrial and marine strata, are composed largely of limestone or dolomite but also include sandstones and shales, including dark graptolite shales. Silurian marine strata commonly contain a large range of invertebrate fossil fauna, including brachiopods, corals, and crinoids (echinoderms). The only known Silurian vertebrates are primitive fish (Agnatha and Placodermi).

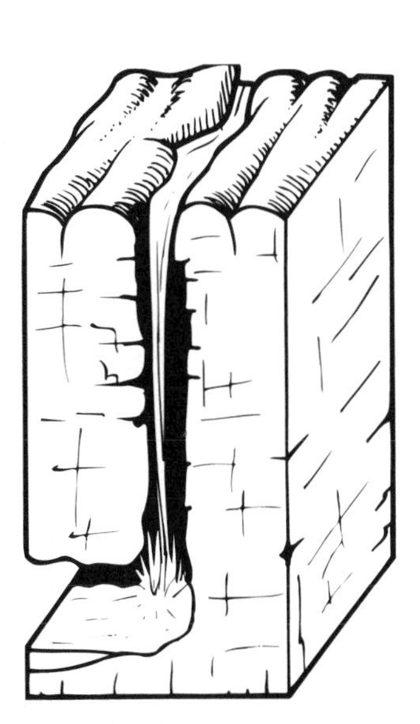

Sinkhole

sinkhole or **doline** A type of hollow in karst limestone regions. Sinkholes form where groundwater dissolves bedrock. *See also* karst.

Siphonaptera The order of bloodsucking insects known as the fleas.

slack Dune slack. A hollow between rows of old and new dunes migrating inland from a sandy shore.

slate A metamorphic rock with flaky minerals aligned in layers by pressure when formed from fine-grained mudstone, volcanic ash, etc. Slate splits readily into sheets used for roofing and flooring.

sleet Precipitation in the form of frozen raindrops or partly melted and then refrozen snow.

slip face The steep advancing side of a sand dune.

slip-off slope The gentle slope of the riverbank on the inside bend of a meander, where sediments may accumulate.

slope A land surface at an angle to the horizontal. Erosion can make a slope retreat or make it less steep, or do both.

slump *See* rotational slippage.

smog Air pollution caused by combined smoke and fog, or (photochemical smog) by sunlight acting on vehicles' exhaust gases.

snow Precipitation in the form of feathery ice crystals that often stick together as snowflakes. The density of snow is increased by packing, melting, and refreezing. In glaciers the density of snow approaches that of pure ice.

sodium A soft silvery white metal, the sixth most plentiful element in the Earth's crust. It occurs in natural compounds and forms nearly one-third of the substances dissolved in seawater.

soil A mixture of mineral fragments, organic matter, air, and water that forms the surface layer of most land and supports the growth of plants. *See also* humus.

soil air The atmosphere within the pore spaces of the soil, consisting of the same gases as the free atmosphere but in different proportions.

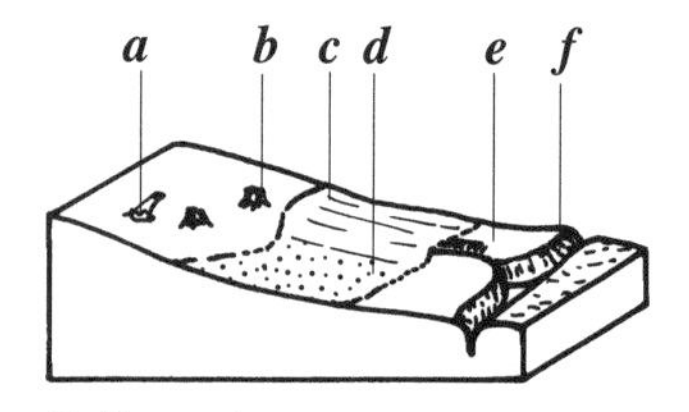

Soil erosion
a *Deforestation*
b *Overgrazing*
c *Downslope plowing*
d *Monoculture*
e *Wind erosion*
f *Flood erosion*

soil association A group of soils that are characteristic of a particular geographic area. A more precise mapping of soils within an area is called a soil complex.

soil conservation All the activities concerned with the protection of soil from erosion, pollution, or mineral deprivation.

soil erosion The wearing away of soil, usually by wind or water. Soil erosion can be severe in lands where overgrazing or removal of vegetation by overplowing destroys the soil structure.

soil formation The weathering, humidification, and other processes by which rocks are reduced in particle size and altered in mineral content, and sometimes acquire organic material, so as to form soil.

soil horizon Any layer in the soil profile distinguishable by mineral of organic content.

soil profile The various layers of soil, from the surface to the underlying rock, as seen in a vertical section.

soil structure The form – aggregate, crumb, granule, or prism – of the soil.

soil-water zone The zone between the soil surface and the water table in which water tends to be held by soil particles.

solar energy (1) The Sun's energy, radiated as heat, visible light, ultraviolet light, and other parts of the electromagnetic spectrum.
(2) Some of this energy is collected, converted, and exploited by solar panels, solar collectors, or solar cells to heat water or air or to generate electricity.

solar radiation Electromagnetic energy over a range of frequencies radiated from the Sun. It includes heat, visible light, ultraviolet and infrared radiation, X rays, and radio waves.

solar system The Sun and its orbiting collection of planets, moons, comets, and asteroids.

solar wind A stream of energetic particles emitted by the Sun.

solifluction In permafrost regions, a process by which soil lubricated by summer meltwater slides downhill on the permanently frozen ground beneath it.

solstice Either of two days in the year when the Sun reaches its highest point in the sky, north or south of the equator. The Northern Hemisphere's summer solstice, and longest day, occurs about June 21, when the midday Sun is above the tropic of Cancer, north of the equator. The Northern Hemisphere's winter solstice, and shortest day, is about December 22, when the Sun is above the tropic of Capricorn, south of the equator.

solution (1) Two mixed substances that cannot be separated by filtering or other mechanical means, for instance, salts dissolved in water.

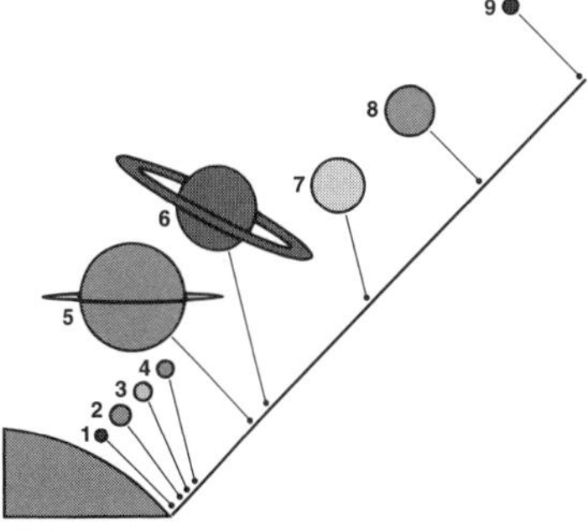

Solar system
1 Mercury
2 Venus
3 Earth
4 Mars
5 Jupiter
6 Saturn
7 Uranus
8 Neptune
9 Pluto

(2) Chemical weathering of rock where water dissolves minerals, for instance, rainwater containing dissolved carbon dioxide that forms weak carbonic acid that dissolves limestone.

source The starting point of a river. Most rivers stem from a spring, but some emerge from a lake, glacier, or underground stream.

South Pole (1) The geographic South Pole, 90° south; an imaginary point in Antarctica at the southern end of the Earth's axis.
(2) Magnetic south pole; a variable point in the Antarctic to which one end of a compass needle points. At the pole itself the needle tries to point downward.

Southern Hemisphere The half of the Earth south of the equator.

space probes Rocket-powered instrument packages sent into space to explore the solar system.

speciation The process of evolution of new species of living things from previous species by spontaneous genetic change (mutation) and natural selection.

sphere Any ball-shaped object.

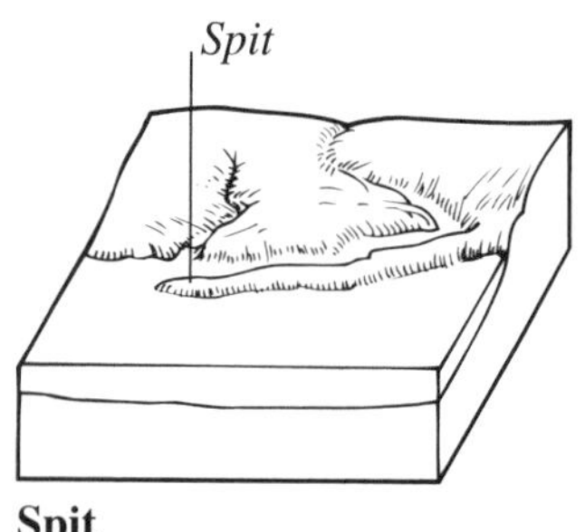

Spit

spherulite A spherical aggregate, about 0.5 of an inch (1 cm) in diameter, of fine, radial, needle-like crystals resulting from the devitrification of a glass.

spit A low strip of sand or shingle with one end joined to land and the other extending into the sea or across a bay.

spoil Waste material removed by mining or quarrying and often piled up to form spoil heaps.

sponges The members of the phylum Porifera, most of which are marine animals, but a few of which live in fresh water.

spreading ridge A submarine mountain chain built by magma that rises to plug a widening gap between two diverging lithospheric plates.

spring (1) The season between winter and summer in temperate latitudes.
(2) Groundwater escaping at the surface, as where a water-saturated rock layer outcrops on a hillside above a layer

of impermeable rock. More generally, a point at which the water table reaches the Earth's surface, thereby producing a usually constant flow of water from the ground.

spring tide The maximum high or low tide that occurs on most coasts twice a month.

spur A ridge jutting out to one side of a mountain or hill. *See also* truncated spur.

squall A fierce wind, often with an intense burst of rain or hail. Squalls arise suddenly but soon die out.

Squamata The main order of reptiles, containing the snakes and the lizards. Nearly 5,000 species have been described, about half of them snakes.

stable platform An eroded shield of ancient rocks overlain by younger ones. *See* shield.

stack An offshore pillar of rock separated from the nearby coast by wave action removing the land between.

stage Rocks formed during an age (a geologic time unit smaller than an epoch).

stalactite A "stone icicle" hanging from a limestone cave roof. It forms from calcium carbonate deposited by water evaporating as it drips from the roof. Stalactites tend to be long and thin, and to have hollow cores. The water moves down the core and precipitates at the bottom, slowly extending the length without closing off the hollow core.

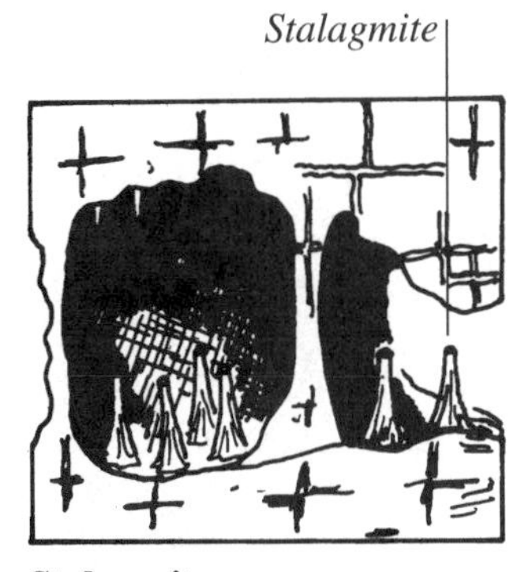

Stalagmite

stalagmite A calcium carbonate column rising from a limestone cave floor. It forms in a similar way to a stalactite and is commonly found immediately below a stalactite. Stalagmites are thicker and shorter than stalactites and do not have a central hollow core. In many cases stalactites and stalagmites join up to form elaborate natural wet columns.

star A vast ball of gas, glowing with radiant energy produced by nuclear fusion taking place in its core.

steam Water in the form of an invisible gas at a temperature above the point at which it boils. As steam cools, condensation produces a white vapor of visible droplets.

steel Iron hardened by the addition of carbon.

Stevenson screen A louvered, box-shaped container on legs, containing weather-measuring instruments. It protects from sunshine, winds, rain, and heat reflected from the ground or buildings.

stock A dome-shaped mass of igneous rock intruded into older rocks; similar to a batholith but smaller.

stone patterns *See* patterned ground.

stope A steplike underground excavation for extracting ore.

storm (1) A severe disturbance of the weather, with fierce winds. (2) Wind of force 11 on the Beaufort scale.

storm beach Boulders and other stones hurled high up on a shore by storm waves.

strait A narrow strip of sea linking two large areas of sea.

stratification The way in which sedimentary rocks form into layers by sedimentation and other means. In general, the lower strata are the older rocks and the details of the stratification can be used to assess relative ages. The fossil record also provides important evidence of age as particular fossils are always associated with particular strata, even if these have been displaced from their original position relative to other strata.

stratigraphic nomenclature A system of naming used by geologists when describing and classifying the rock record of the geologic history of the Earth. The system is based mainly on time, as indicated by the fossils contained in the rocks, but also on the characteristics of the rocks. In general, the method assumes that the lower rocks were the first to be deposited and are the older, and the names for the periods and their corresponding rock formations may be related to the areas where these formations were first described, or to ancient tribes once occupying the areas. The Jurassic, for instance, is named for the Jura Mountains, Permian for Perm in Russia,

Devonian for Devon, Cambrian for Wales, Silurian for the ancient British Silure tribe, and the Ordovician for the ancient Celtic tribe in North Wales.

stratigraphy The branch of geology concerned with the study of stratified rocks, and the history and arrangement of rock strata. Stratified rocks are not necessarily sedimentary rocks, and stratigraphy is also concerned with layered igneous rocks, such as lavas and tuffs, and with metamorphic rocks that were formed from sedimentary or volcanic rocks. Stratigraphy is of great practical and commercial importance, especially in relation to oil exploration.

stratocumulus Layered low clouds that cover the sky.

stratopause A level in the Earth's atmosphere marking the top of the stratosphere and the bottom of the mesosphere.

stratosphere The stable layer of atmosphere immediately above the troposphere that extends from about 6–30 miles (10 –50 km) above the Earth's surface. Its temperature increases with altitude but only to a maximum of the freezing point of water. Ozone in the upper stratosphere is effective in limiting the intensity of ultraviolet radiation from the Sun.

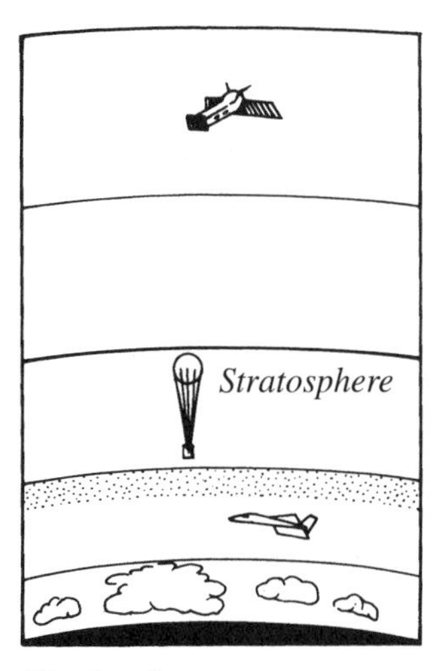

Stratosphere

stratum (pl. **strata**) A sedimentary rock layer.

stratus (pl. **strati**) A low, gray, sheet of cloud. It may produce drizzle but is often thin enough for the Sun to show through it.

stream Water flowing through a channel. Streams include immense rivers and rills small enough to step across.

Strigiformes The order of birds containing the owls (true owls and barn owls).

strike (1) Compass direction of a horizontal line at right angles to the dip of a layer of rocks or a fault.
(2) The discovery of an economically valuable deposit of a mineral.

structural geology The branch of geology concerned with description and analysis of the forms of rock bodies and with how these came about. Primary structural features are those that appeared during the formation of rocks; secondary features are those that were imposed after formation. Structural geology is concerned mainly with secondary features. *See also* tectonics.

structure The relationship, such as folding or bedding, between grain aggregates in a rock.

subduction Literally, a leading under, subduction is an important element in plate tectonics involving the sliding of a lithospheric plate downward at an angle so as to pass under unmoving lithosphere.

subduction zone The zone, set at a descending angle to the Earth's surface, down which the leading edge of a plate of the lithosphere passes so that it moves below the level of other plates. This occurs mainly below oceanic trenches in a deep part of an ocean floor.

subglacial moraine Rock debris eroded by a glacier and lying on the rock floor beneath it.

subglacial stream A stream of meltwater flowing under a glacier.

subhedral Describing crystals in an igneous rock that show some suggestions of regular crystal faces.

submerged coastline Coast where a rise in sea level or a drop in land level has drowned valleys or a coastal plain. Features include fjords, rias, and estuaries.

subsidence (1) Sinking of land, with various possible causes. (2) Sinking of high-pressure air, as in the so-called horse latitudes.

subsoil Weathered material below the topsoil. It holds less humus and fewer roots than topsoil.

subtropical zones Latitudes between the tropical and temperate zones. They lie about 25–35° north and 25–35° south of the equator.

sulfide A compound of sulfur with another element, usually a metal. Metallic sulfides are major sources of ores.

summer In temperate regions, a season between spring and autumn.

Sun The star at the center of the solar system. Around it revolve planets, moons, comets, and asteroids. Its mass is 750 times that of all these, and thermonuclear fusion reactions in its interior make it by far the solar system's brightest, hottest object.

Sun
a Corona
b Prominence
c Core
d Sunspots

sunspots Dark spots on the Sun's surface, cooler than the surrounding photosphere.

supergiant Any of the brightest, largest stars.

supernova The brief but intense explosion of a massive star. Its remnant may form a neutron star or a black hole.

superposition The principle that, in undisturbed layered rocks, the higher a stratum lies, the younger it is.

supratidal The zone just above the high-tide line, which is wetted only by spray or by exceptionally high tides.

surface wave *See* L wave.

surveying The measurement of horizontal distances, differences in elevation, directions, and angles on or near the Earth's surface. These data allow the establishment of locations, and the determination of areas and volumes.

swallowhole A vertical hole down which a stream disappears as it flows underground. Swallowholes occur in some limestone rock. *See also* karst.

swamp Permanently wet, muddy ground supporting plants such as reeds, rushes, and sedges. Trees dominate the mangrove swamps of some low-lying tropical coasts.

swash The surge of water up a beach after a sea wave breaks on a shore.

S wave or **secondary wave** A type of earthquake wave with a sideways shaking effect, the second to reach a seismic observatory. *See also* P wave; L wave.

swell A long, symmetrical undulation of the sea surface.

syenites Medium- to coarse-grained intrusive igneous rocks, usually pale in color, containing up to 90% of feldspars, mainly alkaline, and smaller proportions of quartz, biotite, and hornblende.

Syncline

syncline A downfold in sedimentary rocks, creating a basin or trough in which the strata slope toward the vertical. Compressional forces in the Earth's crust produce synclines from a few yards across to many miles wide. The opposite of an anticline.

synoptic chart *See* weather map.

system The rocks formed during a geological period.

tableland A steep-sided plateau.

Talus

talus or **scree** Bits of rock fallen from a cliff face or steep mountainside and forming a steep slope of loose debris at the base of the cliff or mountain.

tarn A small circular lake in the ice-hollowed floor of a cirque where a rock threshold acts as a dam. *See also* cirque.

taxis Movement of an animal or plant toward or away from a source of stimulation such as light, physical contact, sound, chemical substances, gravity, and air and water currents.

taxon Any group of organisms, such as the contents of a genus, species, or phylum, to which a taxonomic rank or category is applied.

taxonomy The science of classification of living and previously living organisms.

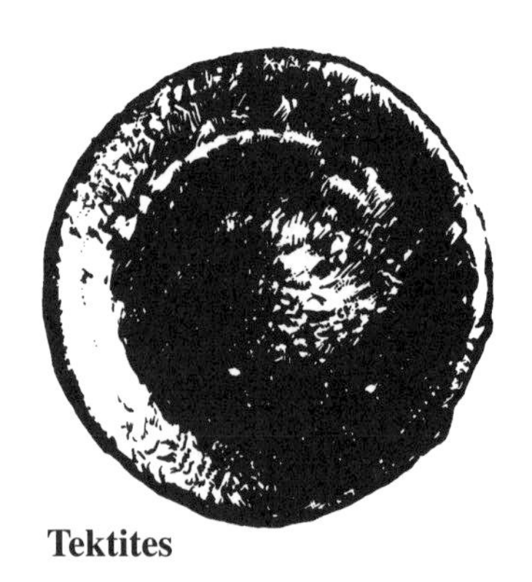

Tektites

tectonics The geological study of the formation of large-scale structures such as mountains. *See* plate tectonics.

tektites Small, black, glassy spheres and "splash-shaped" buttons. They are pieces of meteorites or comets that have been melted on impact with silicon-rich rocks and splashed above the atmosphere and then harden and return, widely dispersed, to Earth.

telescopes Instruments used for detecting and studying distant objects, especially objects in space. (1) Optical telescopes combine parabolic mirrors and lenses to produce a magnified image.

(2) Radio telescopes detect and locate radio waves emitted by stars and galaxies.

temperate zones Latitudes between the Earth's subtropical zones and polar regions.

temperature The degree of heat of the atmosphere or some other body. Atmospheric temperature is usually measured in degrees Celsius (Centigrade) or Fahrenheit (°C or °F).

temperature inversion A condition in which temperature rises with increasing altitude instead of falling. Temperature inversions occur when cold air drains into a valley or warm air moves over cold air or cold ground or water. It can result in rising pollution levels near the ground.

tension Stress caused by forces tending to pull something apart. Tension can produce joints and faults in rocks.

terminal moraine Till (unsorted rubble with clay) dumped along the front of a glacier or ice sheet and left as a long ridge or mound when the ice has retreated.

terrace A shelf-like strip of land in a hillside. Paired river terraces, on opposite sides of a valley, are remains of a floodplain eroded by the downcutting of a rejuvenated river. *See also* rejuvenation.

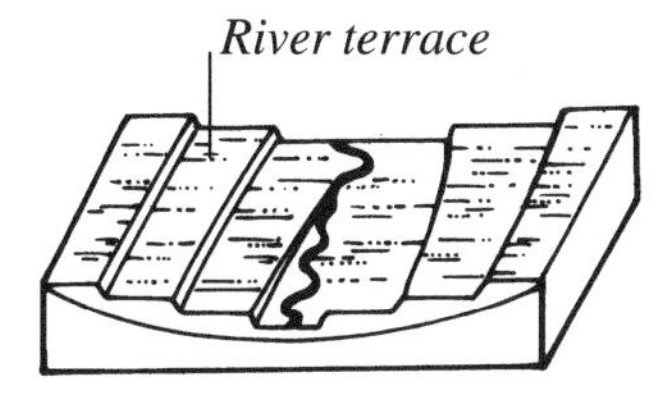

Terrace

terracette An earth step in a hillside caused by soil moving downhill. *See also* creep.

terrane A region bounded by faults that has its own geologic history, stratigraphy, and structural style. Terranes may be microplates, island arcs, or accretionary wedges.

Tertiary period The first period (about 65–2 million years ago) of the Cenozoic era. It extends from the end of the Cretaceous to the beginning of the Quaternary. The Tertiary comprises five epochs: Paleocene, Eocene, Oligocene, Miocene, and Pliocene. The general configuration of the modern continental land masses developed during the Tertiary and the Quaternary. Most groups of mammals now living evolved in Tertiary times.

texture The relationship between the grains in a rock.

thermal metamorphism *See* contact metamorphism.

thermocline A layer of ocean water below which temperatures sharply decline.

thermometer An instrument showing degrees of heat by the amount of expansion or contraction of a heated or cooled liquid or metal, or by variations in how well a metal conducts electricity.

thermosphere The outer layer of the Earth's atmosphere.

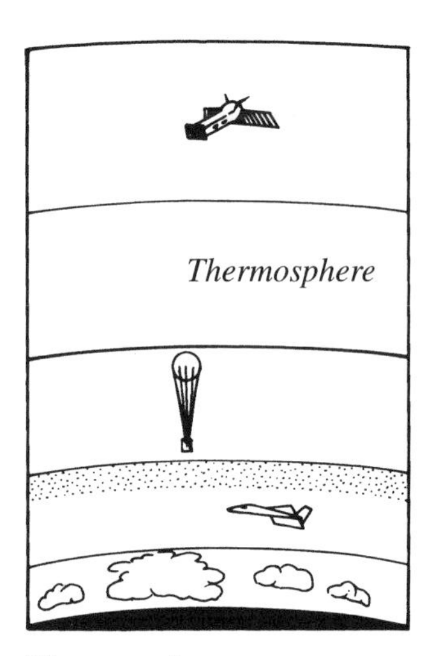

Thermosphere

thunder The bang produced as heat from a flash of lightning makes air around it expand suddenly. The sound of thunder is heard after the flash of lightning is seen because sound travels much more slowly than light.

thunderhead *See* cumulonimbus.

thunderstorm A storm in which towering cumulonimbus clouds generate lightning and thunder, often accompanied by strong winds and a heavy shower of rain or hail.

tidal bore A rare phenomenon in which the rise of a tide in a river is so rapid that water advances as a wall that may be several feet high. A tidal bore occurs when the tide range is unusually high and the river channel narrows rapidly upstream.

tides The regular rise and fall of sea level, mainly due to the Moon's gravitational pull on the Earth.

till or **boulder clay** Sheets of sediment dumped by ice sheets and glaciers. Its ingredients range from boulders to rock flour, clay formed of finely ground rock particles.

tillite Rock formed of consolidated till.

time zones The world's 24 time zones, based on the prime meridian. Clocks are set back one hour with each zone entered to its west. Clocks are set forward an hour with each zone entered to its east. *See also* international date line.

tombolo A spit linking an island to a mainland.

topaz An aluminum silicate mineral valued as a gemstone. It has a hardness of 8 on Mohs' scale and occurs as clear, pale yellow, and (when heated) rose-pink stones.

topographic map A map showing surface features such as hills, rivers, cities, and roads.

topsoil The upper soil layer, usually rich in humus and plant roots.

tor An exposed, weathered mass of heavily jointed rocks crowning a hilltop.

Tor

tornado A fierce whirlwind a few dozen yards across, forming a funnel-shaped cloud. Tornadoes' wind speeds are the highest on Earth. In spring, tornadoes cause severe local damage in southern and midwestern states.

toxic Poisonous.

trace elements Chemical elements in tiny amounts essential to life. They include copper and iodine, found in most soils.

trachytes Fine-grained extrusive igneous rocks with similar mineral content to syenites. They are the extrusive equivalent of the intrusive syenites.

trade winds Constant winds blowing from the subtropics toward the equator. North and south of the equator they are, respectively, the northeast and southeast trade winds.

transform fault A fault at right angles to a spreading ridge and separating two offset segments of the lithospheric plate to one side of the ridge.

transportation (1) In geomorphology, the movement of eroded material by wind, water, or ice.
(2) In human geography, the movement of goods or people from place to place.

travertine Calcium carbonate deposited around a hot spring, as terraces and other formations.

trellised drainage pattern A river system where long parallel streams are joined by short ones flowing at right angles to them.

tremor Trembling of the ground, as produced by a minor earthquake.

trench *See* oceanic trench.

Triassic period The earliest period (about 248–213 million years ago)

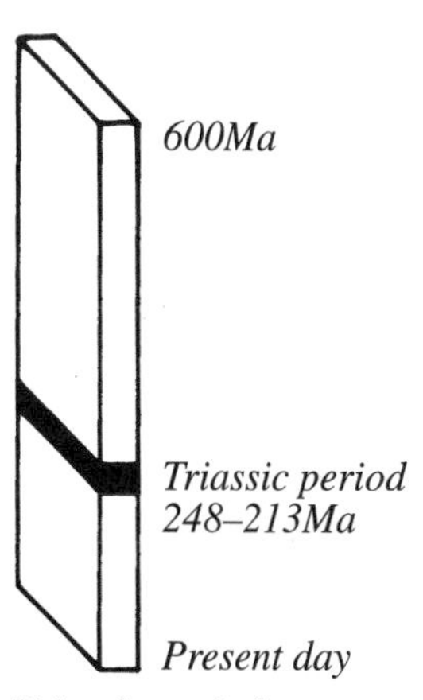

Triassic period

of the Mesozoic era. Dinosaurs, mammals, ammonites, modern corals, various mollusks, and some gymnosperms appeared at this time.

Triassic Table A geological term for the lowest rock system of the Mesozoic era. The name refers to the three-fold facies of strata found first in central Germany but since discovered in other parts of the world. The lowest of these is a non-marine redbed facies; above this is a marine limestone, sandstone, and shale facies; and above this is another non-marine continental facies similar to the lower division.

tributary (1) A stream flowing into a larger stream.
(2) A valley glacier joining a larger glacier.

trilobites Fossil animals of the most primitive arthropod class occurring from the early Cambrian through the Permian. Trilobites were segmented animals with head, thorax, and hind regions and two longitudinal furrows partially dividing the body into three parallel lobes.

triple junction The meeting point of three lithospheric plates.

tropic of Cancer The parallel (line of latitude) 23°north of the equator.

tropic of Capricorn The parallel (line of latitude) 23°south of the equator.

tropics Regions between the tropics of Cancer and Capricorn.

tropism The directional growth of a plant or part of a plant under the influence of an external stimulus such as light, physical contact, or gravity. Tropisms may be positive or negative.

tropopause The boundary between the troposphere and the stratosphere in the atmosphere. The height of the tropopause is about 9.3–11 miles (15–17 km) over the tropics and 6.2 miles (10 km) nearer the Poles. But this varies seasonally and also daily with changes in the weather. It is higher over anticyclones than over depressions.

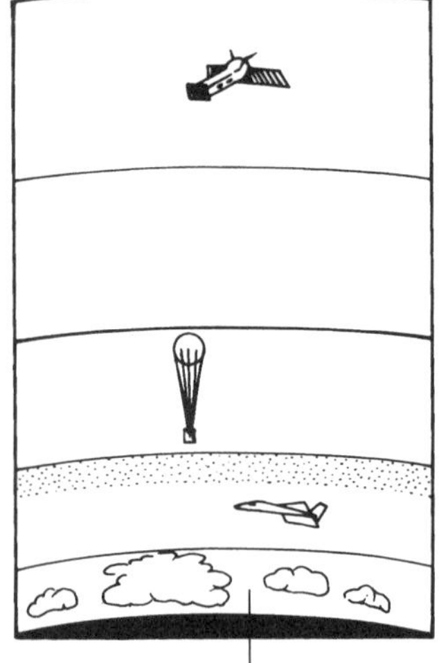

Troposphere

troposphere The lowest level of the atmosphere, extending up to 6–10 miles (10–16 km) above the Earth's surface, to the base of the tropopause. The temperature of the troposphere decreases

GLOSSARY
Triassic Table – troposphere

steadily with altitude at a rate of about 19°F per mile (6.5°C per km), but temperature inversions (increasing with altitude) sometimes occur. Almost all the atmospheric water vapor and suspended aerosols occur in the troposphere. Most atmospheric turbulence and weather systems occur here.

truncated spur A spur with its end lopped off by a valley glacier.

tsunami Seismic sea wave. A high-speed wave set off by an underwater earthquake, landslide, or volcanic eruption. Speeding across an ocean, a tsunami towers on entering shallow water and can swamp coastal settlements. The term is derived from the Japanese *tsu*, "harbor," and *nami*, "wave."

tuff Compacted volcanic ash deposit originating in a pyroclastic process, with grain size of less than 0.1 of an inch (2 mm).

tundra A cold, treeless Arctic zone where the subsoil stays frozen and vegetation consists of mosses, lichens, grasses, and low-growing shrubs. Similar vegetation grows on high mountains, even near the equator.

turbidity currents Rapid movement of a mixture of water and sediment down a slope on the ocean floor. Turbidity currents are often caused by submarine earthquakes and associated with the formation of ocean floor canyons.

typhoon A tropical cyclone of the western Pacific.

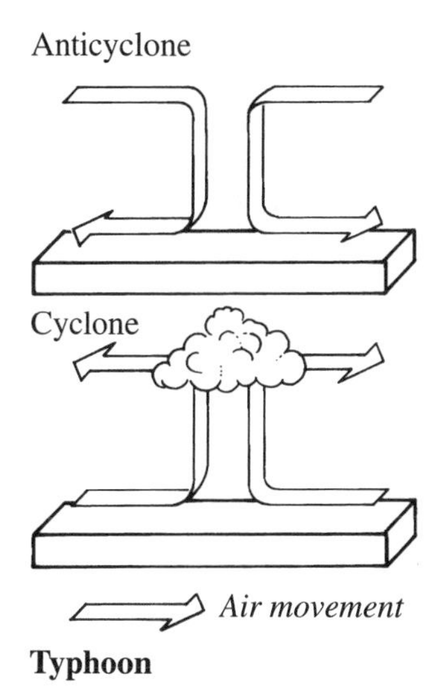

Typhoon

ultrabasic Describing igneous rocks with no free quartz and less than 45% silica. Ultrabasic rocks consist almost entirely of ferromagnesian minerals.

ultramafic Descrbing igneous rocks consisting entirely of dark minerals with no quartz or feldspar.

ultraviolet radiation (**UV**) Electromagnetic radiation with a wavelength just shorter than that of visible light. Atmospheric ozone absorbs much UV light received from the Sun, so helping to reduce its potentially harmful effects upon life.

umbra (1) The conical shadow cast by one heavenly body upon another in an eclipse.
(2) The dark center of a sunspot.

unconformity An interruption in a succession of rock layers, as where one stratum overlies others tilted at a different angle, and shown where erosion has removed an older rock surface.

uniformitarianism The principle that present geological processes are the key to past events in the Earth's history.

universe All known matter and energy, including the whole of the electromagnetic spectrum including light.

upwelling The upward movement of cold water from the depths of the ocean that results when the hotter and lighter surface water is displaced by currents or winds.

uranium A highly radioactive element used in thermal nuclear reactors and atomic bombs.

Uranus The solar system's third largest planet, with a diameter four times that of the Earth. Its rocky core seems to be surrounded by water. A number of moons and rings of stony particles move in orbit around it.

vadose water *See* hydrosphere.

valley A long depression in the Earth's surface, produced by flowing water or ice, or forces compressing or stretching part of the Earth's crust. *See also* rift valley.

Van Allen radiation belts Two doughnut-shaped belts of electrically charged particles surrounding the Earth. Solar activity disrupting the belts can affect radio reception and produce auroras and fluctuations in the Earth's magnetic field.

vaporization *See* evaporation.

variable star A star that varies in brightness as it expands and contracts, or undergoes periodic explosions, or is eclipsed by a binary star partner blocking its light.

varves Paired light-colored coarse, and darker-colored, fine layers of clay sediment laid down, respectively, in summer and winter in lakes near melting glaciers. Varves enable scientists to date some ancient lake sediments.

vein A crack in rock containing a mineral crystallized from a hot solution that once filled the crack.

Venus The second planet from the Sun, almost as large and heavy as the Earth but with a scorching, suffocating atmosphere.

vernal Relating to or happening in spring.

Vertebrata A subphylum of the phylum Chordata of the animal kingdom, containing all animals possessing a backbone (vertebral column). The vertebrates include the fish, amphibia, reptiles, birds, and mammals.

vesicle A small spherical or ellipsoid cavity in an igneous rock caused originally by a gas bubble.

viscosity The resistance offered by any material to its ability to flow.

vitreous Of, resembling, or relating to, glass.

volatiles Substances that pass readily into the gaseous state at ordinary temperatures. Substances with a high vapor pressure.

volcanic glass Natural glass formed by rapidly cooling lava. Volcanic glass is opaque, variously colored red, brown, black, gray, or green, and may be banded. It fractures with smooth shell-shaped concave and convex surfaces (conchoidal fracture). Most natural glasses are chemically equivalent to rhyolite.

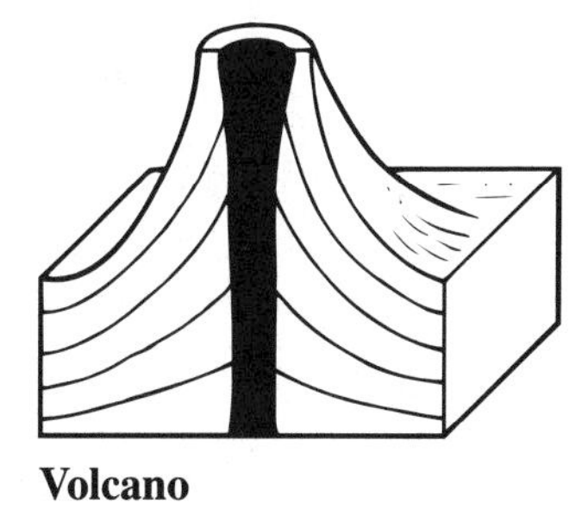

Volcano

volcano A commonly steep-sided, or cone-shaped mountain or hill, formed by the accumulation of hardened magma pierced by a hole or fissure from which lava and/or hot ash and gases erupt from deep underground. Volcanoes are termed active, dormant, or extinct, depending on how often they erupt. Big eruptions produce greater elevation of the volcano or extensive sheets of extrusive igneous rock. Volcanoes and their products are much more prominent on the Earth's Moon, on Mars, and on Jupiter's moons than on Earth.

vug or **vugh** A cavity or irregular opening in a rock that may contain a lining of crystalline minerals.

wadi A normally dry desert watercourse.

warm front The moving boundary between a warm air mass advancing over a cold one, bringing low cloud and rain.

water cycle The circulation of water from sea and land to air and back.

Water evaporates from sea and land, condenses as clouds and falls as rain, hail, sleet, or snow.

water pollution Contamination of rivers, lakes, and seas by fertilizers, pesticides, sewage, and oil or toxic waste.

water table The upper surface of rock saturated by groundwater. A water table's level roughly shadows that of the ground above it. Wet and dry weather make the water table rise and fall.

water vapor Water in air and gaseous in form but below the temperature at which it boils to form steam.

watershed or **drainage basin** Land drained by a river and its tributaries.

wave A disturbance moving through the surface layers of water or land.

wave-cut platform A rock platform extending out to sea just below sea level. It shows where a cliff coast has been cut back and beveled by waves.

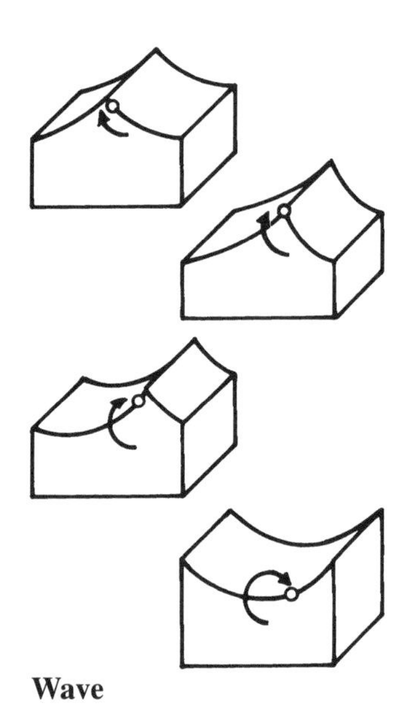
Wave

wave power The energy in sea waves exploited to generate electricity.

waxes Substances containing mixtures of esters of higher fatty acids and long-chain monohydric alcohols.

weather The meteorological state of the atmosphere at a particular geographic locality, especially with regard to atmospheric pressure, temperature, wind speed, cloudiness, humidity, and rainfall. When similar conditions obtain over a wide area, the whole is called a weather system.

weather map or **weather chart** or **synoptic chart** A chart showing the weather at a particular time and place. Weather charts use special symbols to indicate such things as temperatures, pressures, winds, precipitation, and weather fronts. *See also* isobar; isohyet; isotherm.

weathering The decay and breakup of rocks on the Earth's surface by natural chemical and mechanical processes.

weight The force attracting an object to the earth or other heavenly body.

westerlies The prevailing winds of the mid-latitudes.

wet and dry bulb thermometer (psychrometer) An instrument showing the amount of water vapor in air. It features two thermometers: one with a dry bulb and one cooled by a moistened bulb. The temperature difference between the two reveals relative humidity. *See also* hygrometer; humidity.

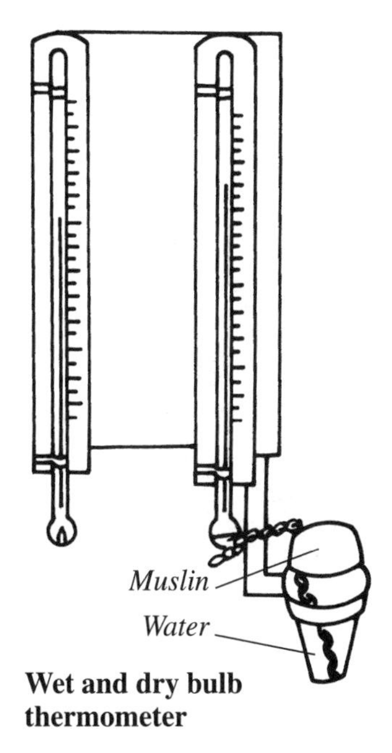

Wet and dry bulb thermometer

wetlands Bogs, marshes, swamps, floodplains, and estuaries. Many have high and varied populations of plants and animals.

white dwarf A faint old star that is far smaller than our Sun, but hotter.

Wilson cycle The hypothesis that ocean basins pass through a series of stages from an initial embryonic stage involving uplift and crust extension through seafloor spreading, basin formation with continental shelves, then the development of instability, sinking to form ocean trenches, shrinkage and continental closure, leading to final destruction. The plate tectonics involved in several of these stages have been identified in various places. This scheme was put forward by Professor John Tuzo Wilson of Toronto University.

wind A horizontal flow of air over the Earth's surface. Winds are described by the directions they come from: for instance, north winds blow from the north. Wind systems help to spread the Sun's heat around the world. *See also* Beaufort scale.

wind measurement The determination of the size, speed, and direction of motion of an air sample.

wind power The kinetic energy in wind exploited by aerogenerators to produce electricity and perform useful work. In the United States, at least 20% of energy is consumed in heating buildings, and in colder climates, heating demand is commonly associated with high wind energy. So wind-generated electricity, combined in some cases with solar heat collectors, can offer a means of reducing fuel oil consumption.

wind vane An instrument with an arrow-shaped arm that rotates to show wind direction.

winter In temperate regions the season between autumn and spring.

winze A steep passage connecting two working levels in a mine.

xenolith A fragment of country rock that is incorporated into intrusive igneous rock.

yardang A narrow rock ridge in a desert. Yardangs develop in upended strata, forming parallel ridges of resistant rock separated by corridors where windblown sand has scoured away weaker rock layers.

yellow dwarf A medium-size star such as the Sun.

Yardang

zenith The point directly overhead in the sky. *Compare* nadir, the diametrically opposite point.

zenithal projections *See* azimuthal projections.

zeolites A large group of minerals of hydrated alumina silicates of the elements sodium, calcium, potassium, and barium. They occur in geodes, hot water cavities (hydrothermal veins), altered igneous rocks, and some sediments.

zeugen Tabular rock formations in deserts. Zeugen develop in horizontal strata where windblown sand widens joints in resistant rock and scours away the softer rock underneath.

zircon A mineral containing the elements zirconium, silicon, and oxygen with various impurities. The pale stones are popular as semiprecious gemstones resembling diamonds.

zodiac A belt of stars that seems to surround the Earth. Its twelve divisions feature two of the apparent star groups called constellations.

zoogeography The scientific study of the geographical distribution of animals.

zooplankton Water creatures that drift or swim only weakly in surface waters, especially tiny creatures that teem in the oceans. *See also* phytoplankton; plankton.

SECTION TWO BIOGRAPHIES

Abbe, Cleveland (1838–1916) American pioneer of meteorology who published more than 300 papers on the subject and became the first official weather forecaster in the United States.

Adams, John Couch (1819–92) English mathematical astronomer and faculty member at Cambridge University. He is best known for his undergraduate astronomical calculations, based on observed irregularities in the orbit of Uranus, which in 1820 predicted the position of an eighth, unknown planet that was discovered in 1846 and called Neptune.

Jean Louis Rodolphe Agassiz

Agassiz, Jean Louis Rodolphe (1807–73) Swiss-born American naturalist and glaciologist. After studying the glacial phenomena of the Alps, he showed that glaciers are not static but move. He was the first to propose the existence of an ice age. Agassiz emphasized the importance of direct personal investigation of natural phenomena. He made the unfortunate pronouncement that there were several distinct species of humans, a view that was used to justify slavery.

Georgius Agricola

Agricola, Georgius (1494–1555) German language teacher, physician, and expert on mining and metallurgy. His important book, *De re metallica* (1556), took him 20 years to write. This was the outstanding text on technology published in the 16th century. Agricola coined the word *fossil* but did not distinguish fossils from other types of rock.

Airy, Sir George Biddell (1801–92) English astronomer and geophysicist who in the 1850s laid a basis for the theory of isostasy – the state of balance in the Earth's crust where continents of light material float on a denser substance into which deep continental "roots" project like the underwater mass of floating icebergs.

von Alberti, Friedrich August (1795–1878) German geologist who in 1834 named the Triassic system, based on a tripartite division of rocks.

Aldrovandi, Ulisse (1522–1605) Italian naturalist who interested himself in biological classification and fossils, and studied gross fetal abnormalities.

Alvarez, Luis Walter (1911–88) American atomic physicist who greatly advanced the field of high-particle physics and, in

1946, developed the proton linear accelerator. He was awarded the Nobel Prize in physics in 1968. With his son, geologist Walter Alvarez; Frank Asaro; and Helen Michel, Alvarez also proposed the theory that the extinction of the dinosaurs was caused by the impact of a large meteorite. Their evidence included the discovery of a larger than normal layer of iridium dated to the appropriate time, about 65 million years ago. The prescence of iridium is characteristic of large meteorite impacts.

Anaxagoras (c. 500–428 BCE) Greek philosopher who was persecuted for believing that the Sun was an incandescent rock. He taught that matter was composed of innumerable tiny particles containing determining qualities.

Anaximander (c. 611–546 BCE) Greek philosopher credited with many imaginative scientific speculations, for example, that living creatures first emerged from slime, and that humans must have developed from some other species that more quickly matured into self-sufficiency.

Anaximander

Anaximenes (fl. c. 500 BCE) Greek philosopher who believed that the origin of all matter was air and that this could be condensed to make various forms of solid matter or liquid. He believed that the Earth was flat.

Arduino, Giovanni (1714–95) Italian geologist and the founder of geology in Italy. Arduino coined the term *Tertiary*, which was later given to a rock system and the corresponding geological period.

Aristotle (384–322 BCE) Greek philosopher who, through his extensive writings, became historically the most influential figure of the ancient world. He covered every field of contemporary knowledge, but in science wrote on physics, biology, medicine, zoology, zoological taxonomy, and psychology. Much of what he wrote about fundamental science was pure imagination and was wrong. Unfortunately, he was accepted as an almost infallible authority, so for almost 2,000 years, scientific thought was misdirected and real progress hindered.

Baade, [Wilhelm Heinrich] Walter (1893–1960) German-born American astronomer whose work, involving new ways of identifying and classifying stars, led him to increase and improve Hubble's values for the size and age of the universe. He also worked on supernovae and on radio stars, and on the Andromeda galaxy.

Sir Joseph Banks

Banks, Sir Joseph (1744–1820) English botanist who accompanied Captain Cook in his voyage around the world from 1768 to 1771 in the ship *Endeavour* and who persuaded King George III to make Kew Gardens in London into a center for botanical research. He caused many tropical fruits, for example, the mango from Bengal, to be imported to the West and was a long-term president of the Royal Society of London.

Barnard, Edward Emerson (1857–1923) American astronomer who gained photographic skills while earning his living in a portrait studio. He became an astronomer and taught while he learned, and reached the rank of professor at the University of Chicago in 1895. While working at the Lick Observatory, he discovered the fifth satellite of Jupiter. He also showed that the dark areas of the Milky Way are actually clouds of obscuring material.

Barrell, Joseph (1869–1919) American geologist who declared that much sedimentary rock did not form under oceans. He coined the terms *lithosphere* and *asthenosphere* and was the first geologist fully to realize the potential of radioactive dating.

Gaspard Bauhin

Bauhin, Gaspard (1560–1624) Swiss botanist and anatomist who in 1623 published a botanical catalog containing more than 6,000 species of plants. The general arrangement was somewhat haphazard and the taxonomy not based on any recognizable principle, but the species are grouped in a manner suggesting modern genera. The book provides a valuable historical account of the state of botany at the time.

Beaufort, Sir Francis (1774–1857) Irish-born British rear-admiral and charter of the seas (hydrographer) best known for introducing the Beaufort scale of wind force. The scale ranges from calm (0) to full hurricane (12). Beaufort joined the navy in 1787, saw active service for more than 20 years, and became official hydrographer to the British Admiralty in 1832.

de Beaumont, Élie (1798–1874) French geologist who published works on the geology of France and on mountain structure.

de la Beche, Sir Henry Thomas (1796–1855) English geologist who organized and directed a geological survey of Britain. While working in Devon, he observed that some rock strata contained plant fossils similar to those of the Carboniferous system but showing no sign of fossils of the Silurian system that preceded the Carboniferous. He inferred that he had found a stratum that came before the Silurian. It was later accepted, however, that another system, called the Devonian, existed on top of the Silurian and under the Carboniferous.

Becher, Johann Joachim (1635–82) German chemist who studied minerals. His *Physica subterranea* (1669) was the first attempt to bring physics and chemistry into close relation.

Johann Joachim Becher

Beebe, Charles William (1887–1962) American naturalist and underwater explorer best remembered for the Beebe bathysphere, a strong, sealed, spherical capsule, fitted with observation ports and light, that could be lowered to great ocean depths for scientific studies. Beebe reached a depth of 3,028 feet.

Bell Burnell, Susan Jocelyn (b. 1943) British astronomer, better known as Jocelyn Bell, who, while still a student, identified an intermittent radio signal from space that was shown to originate outside the solar system and was eventually identified as a rapidly rotating neutron star causing a pulsation. This was the first pulsar to be discovered.

Belon, Pierre (1517–64) French naturalist who wrote works on the natural history of fish and of birds. He appears to have been the first to observe anatomical equivalents (homologs) in the vertebral bones of fish and mammals.

Pierre Belon

Bentham, George (1800–84) British botanist and classifier (taxonomist) who wrote one of the first books on British flora, *The Handbook of the British Flora* (1858). He also wrote extensively on Australian flora. His collection of many thousands of specimens of plants were presented to London's Kew Gardens in 1854.

Bertrand, Marcel-Alexandre (1847–1907) French geologist who showed that the formation of mountain ranges such as the Alps involved massive folding of the Earth's crust.

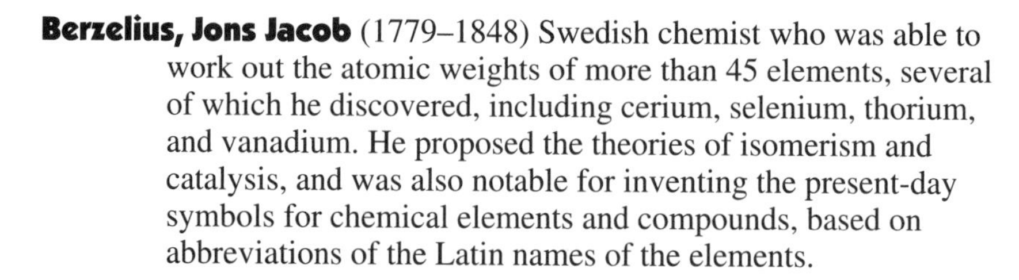

Berzelius, Jons Jacob (1779–1848) Swedish chemist who was able to work out the atomic weights of more than 45 elements, several of which he discovered, including cerium, selenium, thorium, and vanadium. He proposed the theories of isomerism and catalysis, and was also notable for inventing the present-day symbols for chemical elements and compounds, based on abbreviations of the Latin names of the elements.

Jons Jacob Berzelius

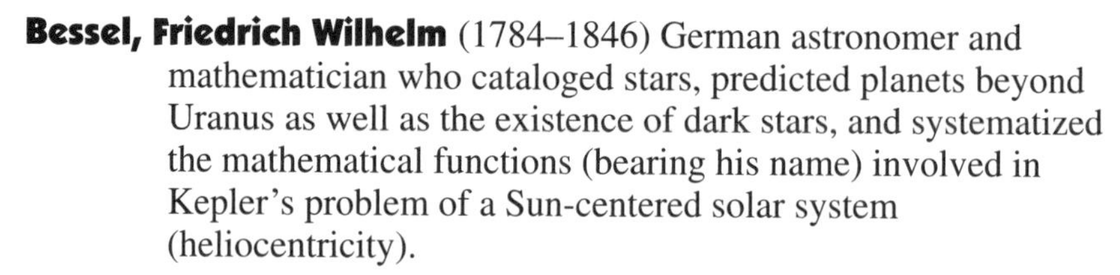

Bessel, Friedrich Wilhelm (1784–1846) German astronomer and mathematician who cataloged stars, predicted planets beyond Uranus as well as the existence of dark stars, and systematized the mathematical functions (bearing his name) involved in Kepler's problem of a Sun-centered solar system (heliocentricity).

Friedrich Wilhelm Bessel

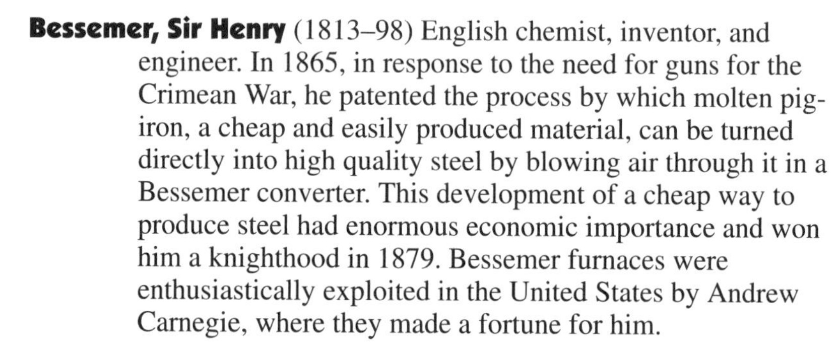

Bessemer, Sir Henry (1813–98) English chemist, inventor, and engineer. In 1865, in response to the need for guns for the Crimean War, he patented the process by which molten pig-iron, a cheap and easily produced material, can be turned directly into high quality steel by blowing air through it in a Bessemer converter. This development of a cheap way to produce steel had enormous economic importance and won him a knighthood in 1879. Bessemer furnaces were enthusiastically exploited in the United States by Andrew Carnegie, where they made a fortune for him.

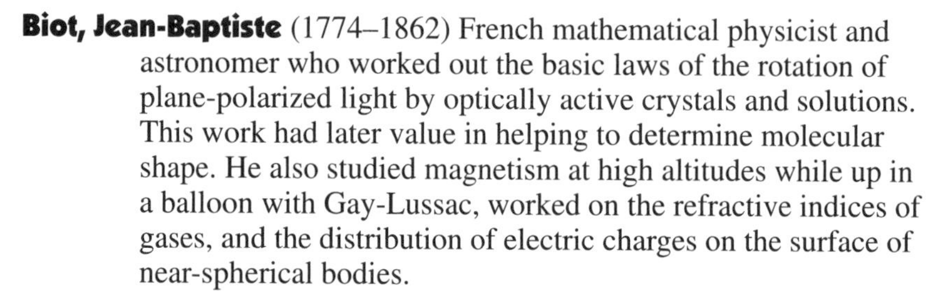

Biot, Jean-Baptiste (1774–1862) French mathematical physicist and astronomer who worked out the basic laws of the rotation of plane-polarized light by optically active crystals and solutions. This work had later value in helping to determine molecular shape. He also studied magnetism at high altitudes while up in a balloon with Gay-Lussac, worked on the refractive indices of gases, and the distribution of electric charges on the surface of near-spherical bodies.

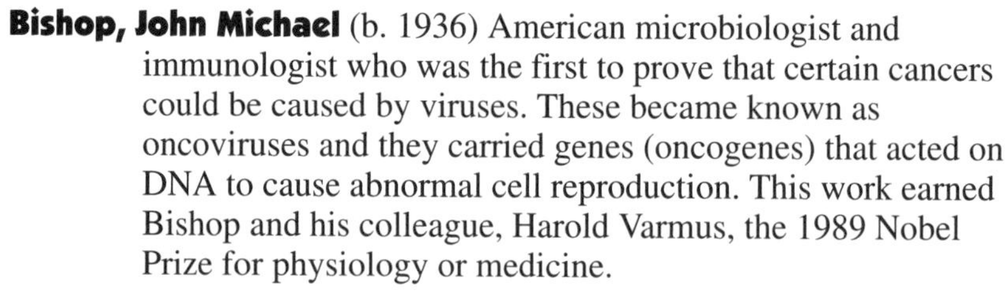

Bishop, John Michael (b. 1936) American microbiologist and immunologist who was the first to prove that certain cancers could be caused by viruses. These became known as oncoviruses and they carried genes (oncogenes) that acted on DNA to cause abnormal cell reproduction. This work earned Bishop and his colleague, Harold Varmus, the 1989 Nobel Prize for physiology or medicine.

Bjerknes, Jacob Aall Bonnevie (1897–1975) Norwegian-born American meteorologist, son of Norwegian physicist Vilhelm Bjerknes. With his father, he formulated the theory of cyclones, on which modern weather forecasting is based.

Jacob Aall Bonnevie Bjerknes

Bjerknes, Vilhelm (1862–1951) Norwegian mathematician and geophysicist who applied hydrodynamics and thermodynamics to his study of weather systems and produced equations for the thermal energy and other characteristics of developing cyclones.

Vilhelm Bjerknes

Bock, Jerome [Hieronymus Tragus] (1498–1554) German botanist who was one of a small group of pioneers of modern scientific botany.

Bode, Johann Elert (1747–1826) German astronomer who became director of the Berlin Observatory. In 1772, he publicized Bode's Law, attempting to describe the mathematical relationship between the planets and the Sun. The law, however, does not hold for the most distant planet, Pluto, and has no theoretical significance.

Borlaug, Norman Ernest (b. 1914) American agronomist and plant breeder who was one of the pioneers of the green revolution in agriculture. He was awarded the 1970 Nobel Peace Prize for his work on breeding improved wheat for India and Mexico.

Bose, Sir Jagadis Chandra (1858–1937) Indian physicist and botanist whose main work was in the polarization and reflection of electric waves and in the investigation of the growth and sensitivity of plants.

Bradley, James (1693–1762) English astronomer who in 1729 published his discovery of the aberration of light – the apparent displacement in the position of a star as a result of the Earth's motion around the Sun. This was the first observational proof of the Copernican hypothesis that the Earth moves around the Sun and not vice versa. In 1748, he discovered that the inclination of the Earth's axis to the ecliptic is not constant. He succeeded Halley as regius professor of astronomy at Greenwich.

Brahe, Tycho or Tyge (1546–1601) Danish astronomer who in 1563 discovered serious errors in existing astronomical tables and in 1572 observed a new star in Cassiopeia, a nova now known as

Tycho's star. From 1576, he developed an accurate catalog of the positions of 777 stars.

Alexandre Brongniart

George Louis Leclerc Buffon

Robert Wilhelm Bunsen

Luther Burbank

Brongniart, Alexandre (1770–1847) French naturalist and geologist who in 1829 introduced the term *Jurassic* for limestones and clays of the Cotswolds in England. Brongniart introduced the idea of using fossils to date and identify disturbed layers of sedimentary rock.

Brown, Robert (1773–1858) Scottish botanist renowned for his investigation into the impregnation of plants. He was the first to note that, in general, living cells contain a central dark-staining mass, and to name it the nucleus. In 1827, he first observed the "Brownian movement" of fine particles in liquid caused by molecular movements.

Brunfels, Otto (1489–1534) German botanist who, with two others, Bock and Fuchs, is considered a pioneer of modern botany. Brunfels also produced accurate and detailed botanical illustrations of real scientific value.

Buffon, Georges Louis Leclerc, Comte de (1707–88) French naturalist who wrote a 44-volume work on natural history that, he hoped, would contain all that was known of the subject at the time. Buffon partly accepted the idea of evolution and realized that all species were, in some way, related.

Bunsen, Robert Wilhelm (1811–99) German experimental chemist and inventor. He developed the gas burner that bears his name and the ice calorimeter. Working with the German physicist Gustav Kirchhoff, he developed the important analytical technique of chemical spectroscopy. Bunsen also discovered the elements cesium and rubidium.

Burbank, Luther (1849–1926) American horticulturist who pioneered the process of improving food plants through grafting, hybridization and other means. He developed the Burbank potato and new varieties of plums and berries. He also developed new flowers, including the Burbank rose and the Shasta daisy.

Calvin, Melvin (1911–97) American biochemist who made notable advances in the understanding of photosynthesis – the processes by which sugars and complex carbohydrates such as starches are synthesized by plants from atmospheric carbon dioxide. Calvin used radioactive carbon tracers to follow the movement of carbon through the complex reactions. He also studied the chemical origin of life and ways to utilize carbon dioxide artificially. Calvin was awarded the Nobel Prize in chemistry in 1961 for his work on photosynthesis.

Camerarius, Rudolph Jacob (1665–1721) German botanist who first identified the male and female reproductive organs in plants.

de Candolle, Augustin Pyrame (1778–1841) Swiss botanist and chemist who introduced the term *taxonomy* for the classification of plants by their morphology rather than physiology, as set out in his *Elementary Theory of Botany* (1813). His new edition of *Flore française* appeared in 1805. He accurately described the relationships between plants and soils, a factor that affects geographic plant distribution. He is remembered in the specific names of more than 300 plants, two genera, and one family.

Cassini, Giovanni Domenico or **Jean Dominique** (1625–1712) Italian-born French astronomer who became professor of astronomy at Bologna in 1560 and first director of the observatory at Paris in 1669. He greatly extended knowledge of the satellites of Saturn and detected the division of its rings, the Sun's parallax, the periods of Jupiter, Mars, and Venus, and zodiacal light.

Henry Cavendish

Cavendish, Henry (1731–1810) English chemist and natural philosopher. In 1760, he discovered the extreme levity of inflammable air and later, at the same time as Scottish inventor James Watt, ascertained that water is the result of the union of two gases. He used the gravitational attraction between bodies of known weight to estimate the weight of the Earth.

Celsius, Anders (1701–44) Swedish astronomer and mathematician best remembered for the Celsius temperature scale proposed by him in 1742 (but with 100° as the melting point of ice and 0° as the boiling point of water). He also measured the relative brightness of stars and calculated the distance of the Earth from the Sun.

Anders Celsius

Cesalpino, Andrea (1519–1603) Italian botanist who wrote a book outlining the principles of botany, including plant structure and an explanation of plant physiology, and proposed a scheme for the classification of plants. This was the first attempt at any formal classification.

Thomas Chrowder Chamberlin

Chamberlin, Thomas Chrowder (1843–1928) American geologist who proposed a "planetesimal" theory of the origin of planets on the basis of solidification of swarming masses of particles. He discovered the origin of windblown deposits (loess), found fossils beneath the Greenland ice sheet, and first proposed dating the ice sheets of the Pleistocene epoch on the geologic time scale. He also contributed to glacial theory and paleoclimatology.

Chambers, Robert (1802–71) Scottish naturalist who believed, correctly, that past geological events should be explained, as far as possible, in terms of processes known still to be occurring. This policy was called *actualism*, and Chambers incorporated it into his biological evolutionary theory. He proposed this theory in his book *Vestiges of the Natural History of Creation* (1844).

Chandrasekhar, Subramanian (1910–95) Indian-born American astrophysicist who showed that when the nuclear fuel of a star is exhausted, gravitational forces begin to pull the material together until it is very dense, atomic electrons are stripped off, and outward pressures from the nuclei halt the compression, thus leaving, in most cases, a white dwarf. Chandrasekhar showed, however, that the greater the mass the smaller the radius, and that a star with a mass greater than 1.4 times the mass of the Sun cannot evolve into a white dwarf. This is known as the Chandrasekhar limit. For this and other advances in astrophysics Chandrasekhar was awarded the 1983 Nobel Prize in physics, with William Fowler.

Sydney Chapman

Chapman, Sydney (1888–1970) English applied mathematician who developed the kinetic theory of gases and made fundamental contributions to geophysics. These included an analysis of geomagnetic variations, a theory of the formation of the ionosphere (sometimes called the Chapman layer), and theories of magnetic storms and lunar tides.

Charpentier, Johann de (1786–1855) Swiss geologist who studied Swiss glaciers and suggested from what he found that they had once been much larger. This view was based on indications of the transport of large rocks and enormous quantities of gravel.

Clairaut, Alexis Claude (1713–65) French geophysicist who established that the Earth is not spherical but is flattened at the poles – it is an oblate spheroid.

Clausius, Rudolf Julius Emanuel (1822–88) German theoretical physicist who greatly advanced the ideas of Nicolas Carnot and James Joule, thereby largely establishing thermodynamics. He cleared up previous difficulties by pointing out that heat cannot pass spontaneously from a cold to a hot body, and furthered the understanding of the kinetic theory of gases. He also promoted the concept of entropy.

Rudolf Julius Emanuel Clausius

Cleve, Per Teodor (1840–1905) Swedish chemist who worked on the rare earths and decided that the "element" didymium, discovered by someone else, was in fact two elements, neodymium and praseodymium. He also discovered holmium and thulium. Ironically, holmium also turned out to be two elements, and in 1886 French chemist Paul Emile Lecoq de Boisbaudran found it was mixed with another new element, dysprosium.

Conybeare, William Daniel (1787–1857) British geologist and Anglican priest who proposed the now discarded theory that geologic changes occur in brief bursts separated by long quiet periods (catastrophism). He also insisted that the evidence provided by fossils could be accounted for by a series of creations of organisms of increasing complexity (progressivism). This idea never really caught on. He was the first to describe the ichthyosaurus.

Edward D. Cope

Cope, Edward D. (1840–97) American paleontologist whose large collection of fossils contained many dinosaur skeletons. Cope wrote extensively on his findings and ideas, and contributed constructively to the evolution debate.

Copernicus, Nicolas (1473–1543) Polish founder of modern astronomy who in 1500 lectured on astronomy in Rome and is believed to have observed an eclipse of the Moon. His *De Revolutionibus*, proving the Sun to be the center of the universe, was

Nicolas Copernicus

completed in 1530 and published just before his death. The timing may have been fortunate, as such beliefs were then held to be heretical and could lead to burning at the stake.

Coriolis, Gaspard Gustave (1792–1843) French physicist who applied his studies of a spinning surface to such phenomena as the way that weather and ocean current patterns differ in the Northern and Southern Hemispheres. He is remembered for the Coriolis effect, which describes the force acting on mobile objects on the earth's surface. He was first to coin the term *kinetic energy*.

Correns, Karl Erich (1864–1933) German botanist who, independently of Mendel, described the laws of inheritance of characteristics. He was also able to show that certain factors are inherited from sources other than from material in the nucleus of the cell. This cytoplasmic inheritance was later shown to be by mitochondrial DNA.

Coster, Dirk (1889–1950) Dutch physicist who, with Hevesy, discovered the element hafnium.

Cousteau, Jacques Yves (1910–97) French oceanographer and inventor who produced a number of underwater devices, including the aqualung and the bathyscaph. He also pioneered underwater photography. As author and filmmaker, he increased public interest in, and awareness of, the underwater world. He also took an active part in the environmental protection movement.

Croll, James (1821–90) Scottish geologist who produced the theory that ice ages are caused by eccentricities in the orbit of the Earth, which cause periodic global temperature drops as the planet swings farther from the heat of the Sun. He investigated these eccentricities and found that they had varied considerably in degree. He postulated that if a major eccentricity coincided with the Earth being farthest from the Sun, an ice age might result.

Cronstedt, Axel F. (1722–65) Swedish mineralogist who in 1751 was the first to isolate nickel. He then demonstrated its magnetic properties. Cronsted wrote an influential work, *Essay towards a System of Mineralogy* (1758), in which he suggested that minerals should be classified by their chemical composition.

Crookes, Sir William (1832–1919) English chemist and physicist who discovered the element thallium and invented the radiometer. Another invention that has had a major social effect was the Crookes tube, by means of which Crookes demonstrated cathode rays (streams of electrons) and showed that they could be deflected by magnets. Cathode-ray tubes are now present in almost every home in the developed world.

Sir William Crookes

Crutzen, Paul (b. 1933) Dutch chemist who, working with F. Sherwood Rowland and Mario Molina, alerted the world to the danger of damage being caused to the ozone layer of the atmosphere, about 9–30 miles up, from artificially produced nitrogen oxides and chlorofluorocarbon (CFC) gases. For this work, the three men were awarded the 1995 Nobel Prize in chemistry.

Culpeper, Nicholas (1616–54) English herbalist, apothecary, and astrologer, whose book *The English Physician* came to be known as "Culpeper's Herbal" and was widely used. It contains little of medical value, and some dangerous prescriptions, but demonstrated that disease could be treated by drugs.

Dana, James Dwight (1813–95) American mineralogist and geologist who classified minerals, coined the term *geosyncline*, studied coral-rock formation, and theorized about the evolution of the Earth's crust. He wrote the first standard reference books in geology and mineralogy.

Daniell, John Frederic (1790–1845) English chemist and meteorologist who invented a hygrometer (1820), a pyrometer (1830), and the Daniell electric cell (1836). His *Introduction to Chemical Philosophy* was published in 1839. Daniell also interested himself in the weather and used his hygrometer to investigate atmospheric humidity.

Dansgaard, Willi (b. 1922) Danish meteorologist who used isotope dating methods to investigate climate in the last 100,000 years. His studies with oxygen isotopes confirmed the most recent ice age, which ended 10,000 years ago, and demonstrated the weather cycle of the last 1,000 years.

Darwin, Charles Robert (1809–82) English naturalist who was the originator, with Alfred Russel Wallace, of the theory of

Charles Robert Darwin

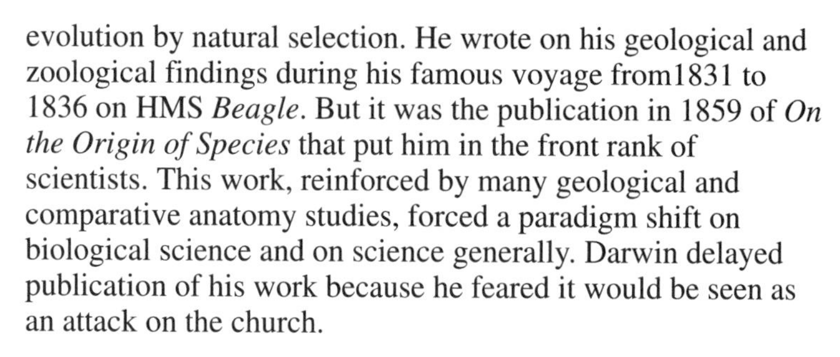
evolution by natural selection. He wrote on his geological and zoological findings during his famous voyage from1831 to 1836 on HMS *Beagle*. But it was the publication in 1859 of *On the Origin of Species* that put him in the front rank of scientists. This work, reinforced by many geological and comparative anatomy studies, forced a paradigm shift on biological science and on science generally. Darwin delayed publication of his work because he feared it would be seen as an attack on the church.

Erasmus Darwin

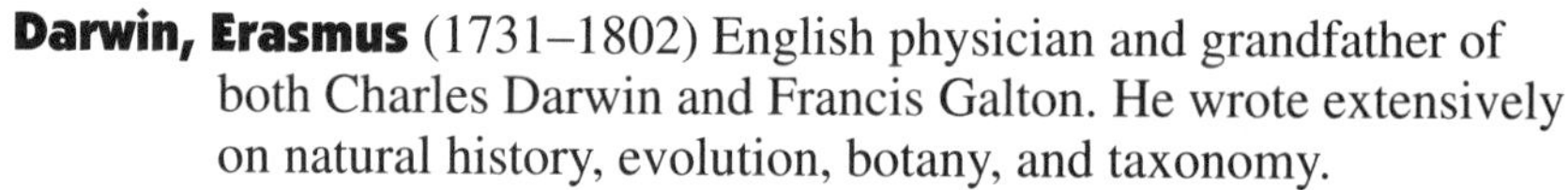
Darwin, Erasmus (1731–1802) English physician and grandfather of both Charles Darwin and Francis Galton. He wrote extensively on natural history, evolution, botany, and taxonomy.

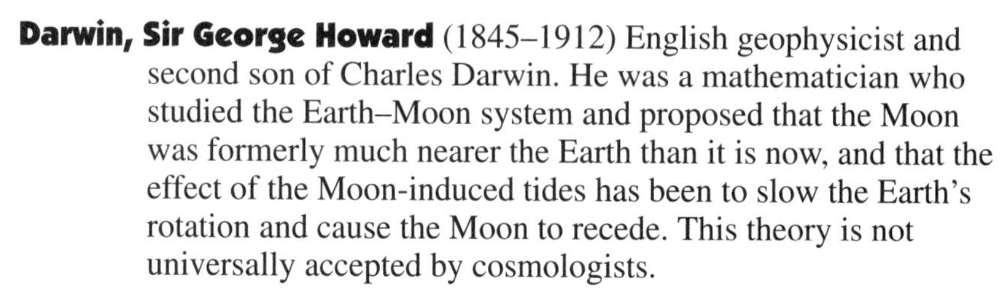
Darwin, Sir George Howard (1845–1912) English geophysicist and second son of Charles Darwin. He was a mathematician who studied the Earth–Moon system and proposed that the Moon was formerly much nearer the Earth than it is now, and that the effect of the Moon-induced tides has been to slow the Earth's rotation and cause the Moon to recede. This theory is not universally accepted by cosmologists.

Daubrée, Gabriel August (1814–96) French geologist and mineralogist who wrote widely on these topics and made some important studies of meteorites. His principal activity was to try to reproduce geological events in the laboratory and he concluded, correctly, that the Earth has an iron core.

Davis, William Morris (1850–1934) American geographer and geologist and founder of geomorphology (scientific landform studies) who introduced the term *peneplain* to describe rolling lowland and was the first to formulate the doctrine of the "cycle of erosion."

Sir Humphry Davy

Davy, Sir Humphry (1778–1829) English chemist and proponent of science who, through his experiments, discovered the new metals potassium, sodium, barium, strontium, calcium, and magnesium. In 1803, he began research in agriculture, and, in 1815, he invented a safe lamp for use in gas-filled coal mines. Michael Faraday worked under Davy and incurred his jealousy and the contempt of Lady Davy.

Democritus (c. 460–c. 370 BCE) Greek philosopher who wrote widely on physics, mathematics, and cosmology. He proposed that all matter consisted of a vast number of tiny particles having a number of basic characteristics, the combinations of which accounted for the variety of objects. This was not a new idea.

Descartes, René (1596–1650) French philosopher and mathematician, remembered for the Cartesian coordinates, who developed coordinate geometry, reformed algebraic notation, produced a mistaken theory of planetary motion, and made an absolute distinction between mind and body, which modern neurological research shows is seriously misleading.

René Descartes

Desmarest, Nicolas (1725–1815) French geologist and vulcanologist who showed how the shapes of landforms are gradually produced by slow erosion and the weathering of geological formations. He also proved that columnar basalts are formed by the cooling of molten rock.

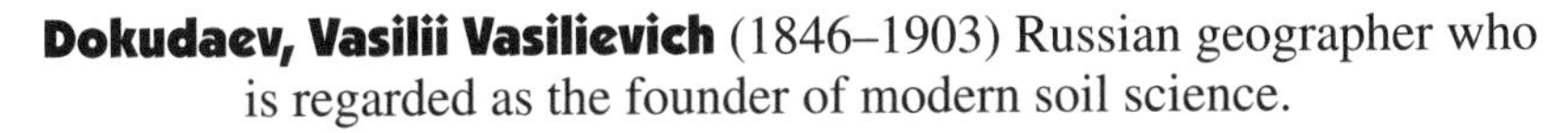

Dokudaev, Vasilii Vasilievich (1846–1903) Russian geographer who is regarded as the founder of modern soil science.

Dorn, Friedrich Ernst (1848–1916) German chemist who discovered the chemically almost inert, but medically dangerous, radioactive gaseous element radon, a noble gas, and showed that it arises as a decay product of radium.

Douglass, Andrew Ellicott (1867–1962) American astronomer and pioneer of dendrochronology – the study of tree rings to date prior climatic events. He found a number of old trees showing six thin rings, indicating a severe local drought at the end of the 13th century. The technique has since been extended back to about 5000 BCE.

Drake, Edwin Laurentine (1819–80) American businessman who drilled the world's first oil well, in Titusville, Pennsylvania, in 1859. His success, although inspired by the profit motive and not, in itself, scientific, had the effect of greatly encouraging geological studies in the search for more oil.

Dubois, Marie Eugène François Thomas (1858–1940) Dutch anatomist and paleontologist who, in the 1890s in Java, found ancient humanoid fossil remains with ape-like characteristics, named *Pithecanthropus erectus*, or Java man, which he claimed to be the missing link between apes and humans, a view eventually accepted in the 1920s.

DuFay, Charles (1698–1739) French chemist whose main contribution to knowledge was in physics. He studied and described the properties of magnetism, showed how magnetic field strength varied with distance, and described natural magnetism. He was also interested in static electricity and was the first to show that an electric charge could be positive or negative, that like charges repelled each other, and that unlike charges are mutually attractive. These discoveries were to have great relevance later to the study of the Earth, its magnetic fields, its geology, and forms.

Du Toit, Alexander Logie (1878–1949) South African geologist who studied the geology of the south of Africa and produced detailed geological accounts. He was one of the earliest geologists to support, in the early years of the 20th century, Wegener's theory of continental drift at a time when most of his colleagues were rejecting it as ridiculous.

Clarence Edward Dutton

Dutton, Clarence Edward (1841–1912) American geologist, seismologist, and vulcanologist who advanced and named the theory of isostasy – the way in which the lithosphere floats on the semisolid asthenosphere beneath it, creating a balance between the heights of the continents and the ocean depths. He also surveyed the structure of the Rocky Mountain region. His studies of volcanoes and earthquakes led him to conclude, correctly, that the rocks that make up the continents are lighter than oceanic rock.

Sir Arthur Stanley Eddington

Eddington, Sir Arthur Stanley (1882–1944) English astrophysicist whose work on what stars are made of had a major impact on astrophysics and cosmology. He established the law of the relationship of the mass of a star to its luminosity (brightness), and showed that the inward gravitational pressures in a star must be, during the star's active life, in exact balance with the outward forces caused by gas pressure and radiation.

Eichler, August Wilhelm (1839–87) German botanist who was one of the outstanding systematic and morphological botanists of his time. He studied flower symmetry and worked on the taxonomy of the higher plants.

Ekman, Vagn Walfrid (1874–1954) Swedish oceanographer who studied the direction of ocean currents, as revealed by drifting

Arctic sea ice, and compared them to those of the prevailing winds. He explained the 45 degree discrepancy in their directions as an effect of the Earth's rotation on wind direction, the Coriolis effect. Ekman was one of the founders of modern oceanography and invented many instruments to measure various ocean parameters.

D'Elhuyar, Don Fausto (1755–1833) Spanish mineralogist who discovered the element tungsten.

Elsasser, Walter Maurice (1904–91) German-born American geophysicist who was the first to propose that the Earth's magnetic field is due to the electric currents flowing in the molten iron outer part of the core of the Earth. He also studied the slow alterations (secular changes) in the magnetic field of the Earth.

Elton, Charles Sutherland (1900–91) English ecologist and author of classic books on animal ecology. His work on animal communities led to a recognition of the ability of many animals to counter environmental disadvantages by a change of habitat, and to the use of the concepts of food chain and niche.

Emiliani, Cesare (b. 1922) Italian-born American geologist who used oxygen isotope analysis techniques to study Pleistocene fossils and prove that there had been seven ice ages in the Pleistocene rather than four, as had been previously believed.

Empedocles (fl. 450 BCE) Greek philosopher who wrote a poem called *On Nature* in which he claimed that everything was made of the four elements, earth, air, fire, and water, which either combined or repelled each other. This idea was to hold up the advancement of chemistry for 2,000 years. Empedocles is said to have jumped into the crater of the volcano Mount Etna in an attempt to prove that he was immortal.

Epicurus (341–270 BCE) Greek philosopher who proposed that everything was made from atoms – particles so small that they cannot be subdivided further. The Roman poet Lucretius accepted and described the atomic theory of Epicurus. The Greek word *atom* means "unable to be cut."

Eratosthenes (c. 276–c. 194 BCE) Greek geographer and mathematician who produced the first world map showing lines of latitude and longitude. He made a calculation of the Earth's circumference accurate to 10% and showed a way of calculating prime numbers.

Eskola, Pentti Elias (1883–1964) Finnish geologist who specialized in metamorphic formations – preexisting rock that has been modified by heat, pressure, or chemical action. In 1915, he asserted that in such rock that has reached chemical equilibrium, the mineral composition is controlled only by the chemical composition. Eskola recognized eight types of metamorphic formation.

Espy, James Pollard (1785–1860) American meteorologist who was the first to suggest that clouds are formed when moist air rises and cools, loses latent heat, and condenses its water to a vapor.

William Maurice Ewing

Ewing, William Maurice (1906–74) American physicist and marine geologist who performed extensive geophysical explorations of the oceans. He measured the thickness of the Earth's crust under the oceans and showed it to average about 4 miles, as compared with 22–25 miles under the continents. He investigated the transmission of sound waves through water and their reflection from the seabed, and produced data that led eventually to the general acceptance of the theory of plate tectonics.

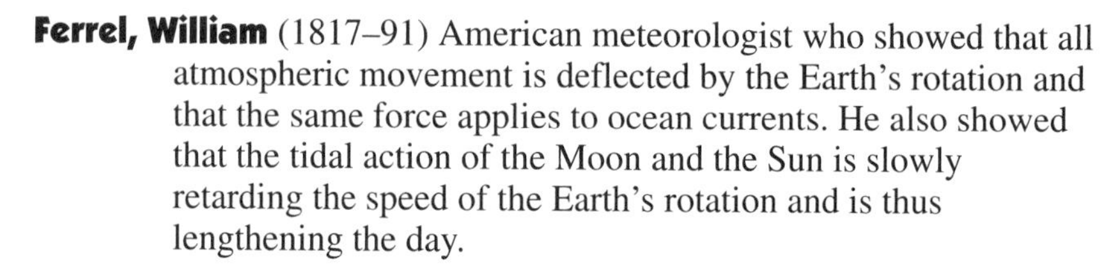

Ferrel, William (1817–91) American meteorologist who showed that all atmospheric movement is deflected by the Earth's rotation and that the same force applies to ocean currents. He also showed that the tidal action of the Moon and the Sun is slowly retarding the speed of the Earth's rotation and is thus lengthening the day.

Fibonacci, Leonardo (c. 1170–c. 1250) Italian mathematician who made many contributions to practical mathematics and number theory but who is best remembered for the Fibonacci series of numbers 1, 1, 2, 3, 5, 8, 13, . . . in which each number is the sum of the previous two numbers. This series describes many patterns in nature.

Fitzroy, Robert (1805–65) English naval captain, hydrographer, and meteorologist who was in command of the *Beagle* on its voyage when Charles Darwin formulated his ideas of evolution by natural selection. Fitzroy designed the type of barometer that bears his name, set up weather observation stations, and initiated the practice of publishing storm warnings.

Flamsteed, John (1646–1719) English clergyman who was appointed first astronomer royal of England in 1675. The following year, he began observations that initiated modern practical astronomy and formed the first accurate catalog of the fixed stars.

John Flamsteed

Forbes, Edward (1815–54) British naturalist who was the first to suggest, correctly, that living organisms might exist deep in the oceans, below the level that can be reached by sunlight.

Franklin, Benjamin (1706–90) American scientist, statesman, and printer who made a number of important contributions to the science of electricity. He is remembered for flying a kite in a thunderstorm to prove that lightning is electrical in nature and showed how buildings could be protected from lightning strikes by metal electrical conductors running down to the ground. He also worked out the course of the Gulf Stream in the Atlantic.

Benjamin Franklin

von Frisch, Karl (1886–1982) Austrian ethologist and zoologist who was a key figure in developing ethology, using field observation of animals combined with ingenious experiments. He showed that forager honeybees communicate information on the location of food in part by the use of coded dances.

Fuchs, Leonhard (1501–66) German botanist and physician who described hundreds of German and other plants, and is remembered in the name of the genus *Fuchsia*. He, together with Brunfels and Bock, was one of the German pioneers of modern botany.

Leonhard Fuchs

Galilei, Galileo, known as Galileo (1564–1642) Italian astronomer, mathematician, and natural philosopher who proposed and proved the theorem that showed that all falling bodies, great or small, descend with equal velocity. He believed, after astronomical research, in the truth of the Copernican heliocentric theory and that the Moon is illuminated by reflection. He produced an improved refracting telescope and was nearly condemned for heresy by the Inquisition. His friend Bruno was burned at the stake for the same beliefs.

Galileo

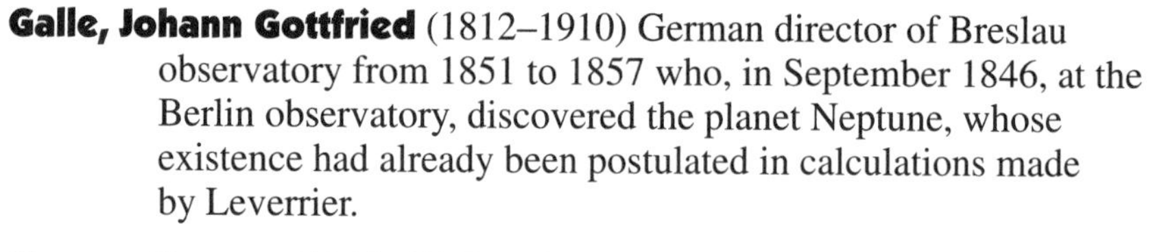

Galle, Johann Gottfried (1812–1910) German director of Breslau observatory from 1851 to 1857 who, in September 1846, at the Berlin observatory, discovered the planet Neptune, whose existence had already been postulated in calculations made by Leverrier.

George Gamov

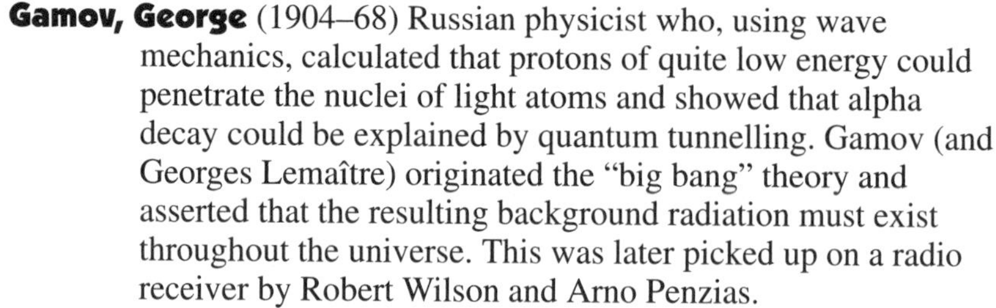

Gamov, George (1904–68) Russian physicist who, using wave mechanics, calculated that protons of quite low energy could penetrate the nuclei of light atoms and showed that alpha decay could be explained by quantum tunnelling. Gamov (and Georges Lemaître) originated the "big bang" theory and asserted that the resulting background radiation must exist throughout the universe. This was later picked up on a radio receiver by Robert Wilson and Arno Penzias.

Gassendi, Pierre (1592–1655) French philosopher and astronomer who studied atomism, acoustics, heat, and thermodynamics, and, in his book on the theory of atoms, *Syntagma philosophicum* (1660), introduced the term *molecule* to indicate the smallest unit of a substance capable of an independent existence.

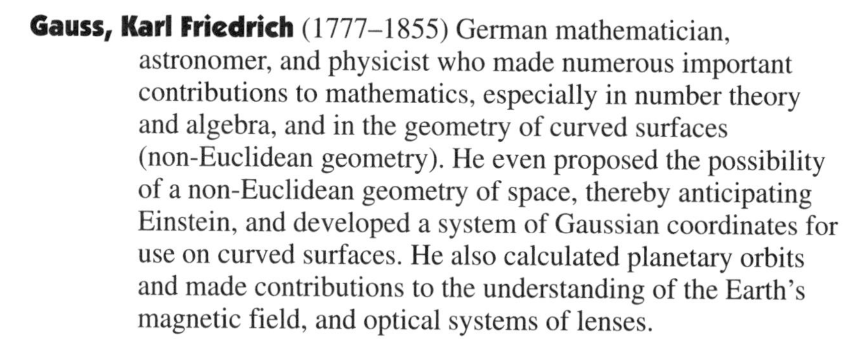

Gauss, Karl Friedrich (1777–1855) German mathematician, astronomer, and physicist who made numerous important contributions to mathematics, especially in number theory and algebra, and in the geometry of curved surfaces (non-Euclidean geometry). He even proposed the possibility of a non-Euclidean geometry of space, thereby anticipating Einstein, and developed a system of Gaussian coordinates for use on curved surfaces. He also calculated planetary orbits and made contributions to the understanding of the Earth's magnetic field, and optical systems of lenses.

Sir Archibald Geikie

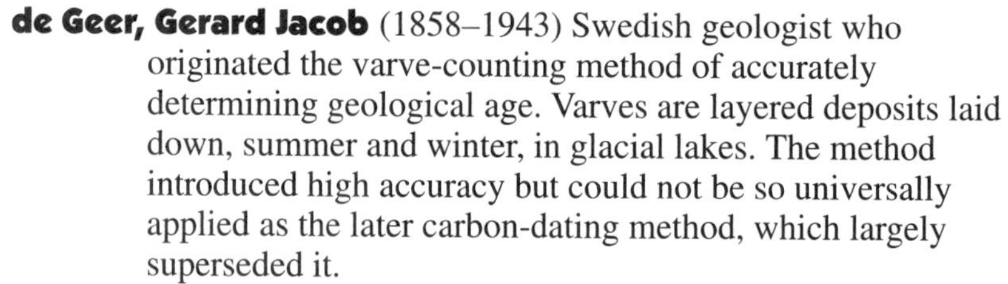

de Geer, Gerard Jacob (1858–1943) Swedish geologist who originated the varve-counting method of accurately determining geological age. Varves are layered deposits laid down, summer and winter, in glacial lakes. The method introduced high accuracy but could not be so universally applied as the later carbon-dating method, which largely superseded it.

Geikie, Sir Archibald (1835–1924) English geologist who over a period of many years directed, successively, both the Scottish and the English Geological Survey. He was especially interested in soil erosion, much of which he attributed to the action of rivers. He wrote a textbook of geology and a number of other works, including *Ancient Volcanoes of Great Britain* (1897).

Gellibrand, Henry (1597–1636) English mathematician and navigator who in 1635 discovered a slow (secular) variation in the angle (the declination) that the compass needle makes with true north. This, unfortunately, put paid to the hope that a consistent declination might be used to indicate geographical longitude. Even more rapid variations were later found.

Gerard, John (1545–1612) British botanist who wrote a book entitled *Herball*, which contained most of the botanical knowledge of his day.

Gesner, Konrad (1516–65) Swiss naturalist who collected fossils and published a book, illustrated with woodcuts, to describe them. He devised an alphabetical scheme of classification of living things and published a compendious catalog of animals called *Historiae animalium* (1551–65).

Gilbert, Grove Karl (1843–1918) American geologist and founder of landform studies (geomorphology) who formulated many of the laws of geological processes. His report on the Henry mountains was the foundation of many modern theories of denudation and river development. He contributed detailed descriptions of river and other geologic processes that became the standards for his time.

Grove Karl Gilbert

Gill, Theodore Nicholas (1837–1914) American fish specialist (ichthyologist) who was an leading taxonomist of his time. His writings greatly advanced the field of ichthyology.

von Goebel, Karl (1855–1932) German botanist and distinguished plant morphologist who wrote *Organographie der Pflanzen* (1898–1901), and founded the botanical institute and gardens in Munich.

von Goethe, Johann Wolfgang (1749–1832) German poet, novelist, lawyer, philosopher, prime minister, physicist, botanist, geologist, and comparative anatomist. Because of his status as an acknowledged genius, Goethe's views on scientific matters commanded a good deal of respect in his time, but his real contribution to the advancement of science was negligible and some of his assertions were seriously misleading. He tried to refute Newton's theory of light and proposed a theory of color vision. Even so, Goethe inspired many people to take up the study of science.

George Brown Goode

Goode, George Brown (1851–96) American fish expert (ichthyologist) and U.S. fish commissioner 1887–88, who wrote *American Fishes* (1888) and *Oceanic Ichthyology* (1895).

Gould, Stephen Jay (b. 1941) American paleontologist who put forward the hypothesis that evolution and new speciation are not continuous processes but occur in fairly sudden jumps separated by long periods in which little or no change occurs. He also proposed that natural selection occurs in genes as well as in whole organisms. He is a highly successful popular science writer.

Asa Gray

Gray, Asa (1810–88) American botanist who, between 1838 and 1842, published the book *Flora of North America*. He also produced *Genera florae Americae Boreali-Orientalis illustrata* (1845–50), *A Free Examination of Darwin's Treatise* (1861), and *Manual of the Botany of the Northern United States* (1848), known as Gray's Manual.

Gregor, William (1761–1817) English chemist whose interest in analyzing local soils led him to the discovery of the element titanium, which has since become an important metal for its light weight and resistance to corrosion.

Grew, Nehemiah (1641–1712) British botanist who carried out some of the earliest research into plant anatomy.

Guettard, Jean Etienne (1715–86) French geologist and mineralogist who studied "weathering" and prepared possibly the first geological map of France. In 1751, he thought he had proved that a number of peaks in the Auvergne were the cones of extinct volcanoes. He later changed his mind.

Gutenberg, Beno (1889–1960) German-born American geophysicist whose research into seismology led him to propose a layer in the Earth's mantle in which seismic waves travel slowly. He measured the radius of the Earth and concluded that it had a liquid core.

Hadley, George (1685–1768) English philosopher and meteorologist who correctly explained the reason why air does not flow directly from north and south to the equator – because the Earth is rotating. Air moving towards the equator has a slower rotational rate than air at the equator, because it is moving in a smaller circle, so will be deflected to a northeasterly in the Northern Hemisphere and a southeasterly in the Southern Hemisphere.

Stephen Hales

Hales, Stephen (1677–1761) English botanist and chemist, founder of plant physiology, whose book *Vegetable Staticks* (1727) was the start of our understanding of vegetable physiology. He was one of the first to use instruments to measure the nutrition and movement of liquids within plants. He also invented machines for ventilating, distilling sea water, and preserving meat.

Hall, Asaph (1829–1907) American astronomer who discovered the Martian moons. He also investigated satellite orbits, planetary perturbations, the solar parallax, and the orbits of double stars. Curiously, the size and orbital periods of Hall's two Martian moons had been accurately described 150 years before by Jonathan Swift in *Gulliver's Travels.*

Hall, Sir James (1761–1832) Scottish geologist who proved the igneous origin of basalt and dolerite rocks by laboratory tests in which he melted and recrystallized minerals. He also showed that molten magma could cause changes in limestone, producing metamorphic rock.

Edmund Halley

Halley, Edmund (1656–1742) English astronomer, physicist, mathematician, seaman, and explorer who made many astronomical discoveries but is best known for correctly predicting the return (in 1758, 1835, and 1910) of the comet that is now named after him. Halley was the first to propose that nebulae are clouds of interstellar gas in which stars are being formed. He was also the first to make a complete observation of the transit of Mercury.

John Harrison

Harrison, John (1693–1776) English horologist and instrument maker who devised a chronometer of such accuracy that it could be used to determine longitude at sea to just over one minute of angle (1/60 of a degree). This clock won the prize of £20,000 that had been offered by the British Admiralty for an accurate chronometer. Harrison and his competitors for the prize became the subject of a best-selling book, *Longitude* (1995), by Dava Sobel.

Hatchett, Charles (c. 1765–1847) English chemist who discovered the metallic element columbium (now niobium), and for whom hatchettine, or hatchettite, a yellowish white semitransparent fossil resin or waxlike hydrocarbon found in South Wales coal, was named.

Hawking, Stephen (b. 1942) British theoretical physicist and professor of mathematics whose work has greatly advanced our knowledge of space-time, black holes – singularities in space-time whose mass is so great that no light can escape – and the quantum theory of gravity. Hawking is best known for his best-selling book, *A Brief History of Time*, and for his continuing to work in spite of suffering from Lou Gehrig's disease (motor neuron disease), which has resulted in his almost complete paralysis.

Hays, James Douglas (b. 1933) American geologist who in 1971 reported evidence from deep-sea piston cores that eight species of radiolaria had become extinct in the last 2.5 million years. Radiolaria are tiny protists that are extremely abundant in the oceans and have played a major role in the geologic history of the Earth as their shells form much of the limestone and chalk sedimentary rocks now present on Earth.

Johannes Baptista van Helmont

Helmont, Johannes Baptista van (1579–1644) Flemish physician, physiologist, and chemist who invented the word *gas*, deriving it from a Greek word for "chaos." He distinguished gases other than air; regarded water as a prime element; believed that digestion was due to "ferments" that converted dead food into living flesh; proposed the medical use of alkalis for excess acidity; and believed in alchemy. His works were published by his son.

Herschel, Caroline [Lucretia] (1750–1848) German astronomer, younger sister of William Herschel, who, although treated as a servant and allowed little or no formal education, became the most celebrated woman astronomer of her time. She discovered eight new comets and several nebulae and clusters of stars, and in 1798 published a radical revision of an existing but unsatisfactory star catalog.

Herschel, Sir John Frederick William (1792–1871) English astronomer who continued and expanded the research of his father, William Herschel, discovering 525 nebulae and clusters. He pioneered celestial photography and carried out studies on photoactive chemicals and the wave theory of light.

Sir William Herschel

Herschel, Sir [Frederick] William (1738–1822) German-born British astronomer who greatly added to knowledge of the solar system, the Milky Way, and nebulae. In 1781, using his own reflecting telescope, he discovered the planet Uranus. He also made a famous catalog of twin stars.

Hertzsprung, Ejnar (1873–1967) Danish astronomer whose important later work on the evolution of stars began with his studies of the relationships between the color and brightness of stars. Hertzsprung also greatly developed Leavitt's method for finding stellar distances, and used it to find the distances of stars outside our own galaxy.

Hess, Harry H. (1906–69) American geologist and geophysicist who described the ocean floors and showed how they are constantly spreading as a result of the upthrust of magma into the mid-ocean ridges.

Hevelius, Johannes (1611–87) German astronomer who cataloged 1,500 stars, discovered four comets, and was one of the first to observe the transit of Mercury. He made a map of the Moon and gave names to many lunar features in his book *Selenographia* (1647).

Hipparchus (c. 180–125 BCE) Greek astronomer and inventor of trigonometry who discovered the precession of the equinoxes and the eccentricity of the Sun's path; determined the length of the solar year; estimated the distances of the Sun and Moon from the Earth; drew up a catalog of 1,080 stars; and used mathematical methods to fix the geographical location of places by longitude and latitude.

Hisinger, Wilhelm (1766–1852) Swedish mineralogist who discovered the element cerium and published a geological map of southern and central Sweden.

Holmes, Arthur (1890–1965) English geologist and geophysicist who put dates to the geological time scale. He determined the ages of rocks by measuring their radioactive constituents, and was an early scientific supporter of Wegener's continental drift theory. He was the first to recognize that the Earth's crust formed (solidified) about 4.56 billion years ago. Holmes published his final time scale in 1959. His book *Principles of Physical Geology* (1944) was highly successful.

Sir Joseph Dalton Hooker

Hooker, Sir Joseph Dalton (1817–1911) English botanist and traveler who went on several expeditions that resulted in works on the flora of New Zealand, Antarctica, and India, as well as *Himalayan Journals* (1854) and the monumental work *Genera Plantarum*. From one trip to the Himalayas, he introduced (1849) the rhododendron to Europe. He, with Thomas Henry Huxley, was a friend of Charles Darwin and strongly supported Darwin's theory of evolution.

Sir Fred Hoyle

Hooker, Sir William Jackson (1785–1865) British botanist and father of Sir Joseph Hooker, who helped to found and was a director of the Royal Botanical Gardens at Kew, London. He was a writer and illustrator of botanical texts, and his private collection was once one of the finest herbariums in Europe.

Horton, Robert Elmer (1875–1945) American engineer, hydrologist, and geomorphologist who described in detail the movement of water after rainfall. This study was the foundation of modem theories on river basin hydrology.

Hoyle, Sir Fred (b. 1915) English mathematician, astronomer, and science fiction writer who proposed a steady-state theory of the universe and who, in a joking but serious criticism of the theory that the universe started in a single point, referred to this, disparagingly, as the "big bang" theory. The joke backfired – the name stuck. He contibuted to astrophysics with reasearch into the age and evolution of stars.

Edwin Powell Hubble

Hubble, Edwin Powell (1889–1953) American lawyer, athlete, astronomer, and cosmologist best known for his measurement, by Doppler red shift in 1929, of the accelerating recession of distant galaxies. The farther away they are, the faster they are receding. This led to the acceptance of the idea of an expanding universe. The first space telescope, which, after some preliminary problems, is now enormously expanding astronomical knowledge, is named after Hubble.

Huggins, Sir William (1824–1910) English astronomer who carried out spectroscopic studies of the physical constitution of stars, planets, comets, and nebulae. He used Doppler shift to measure the radial motion of stars, established that some nebulae are composed of luminous gases, ascertained the luminous properties of certain comets, determined the amount of heat reaching the Earth from some fixed stars, and introduced dry plate photography to astronomy.

von Humboldt, Friedrich Heinrich Alexander (1769–1859) German naturalist, explorer, and scientific observer who made many contributions to Earth sciences. His work on geomagnetism led to the discovery of the magnetic pole. He also identified the Jurassic period of geologic time but is best remembered for his great work of scientific popularization, *Kosmos* (1845–62), which was widely influential on lay scientific thought in his time.

Hutton, James (1726–97) Scottish geologist, physician, and businessman who pioneered uniformitarianism, the belief that forces still at work had caused geological change over vast spans of time. He proved that some rock is the result of the protrusion of molten material from the interior of the Earth (igneous) and that granite is igneous, not sedimentary, as previously thought. Hutton's ideas formed the basis of modern geology. His major work was *Theory of the Earth* (1795).

Sir Julian Sorell Huxley

Huxley, Sir Julian Sorell (1887–1975) English biologist and prolific popular science writer, grandson of Thomas Henry Huxley, who did much to spread biological knowledge among the lay public of his day. He studied the development of organisms and promoted classification, using geographical distribution, as well as physiology and other factors, to determine species.

Pierre-Jules-César Janssen

Janssen, Pierre-Jules-César (1824–1907) French astronomer who showed that the surface of the Sun has a fine granular structure and that apparent prominences on the surface, seen during eclipses, are gaseous. This was demonstrated by spectroscopic analysis. He also developed a method of studying the Sun's prominences, other than during periods of total eclipse.

Jeffreys, Sir Harold (1891–1989) British geophysicist and astronomer whose 1924 book on the origins and formation of the Earth was widely influential. Unfortunately, he rejected Wegener's theory of continental drift, and his views carried so much weight that Wegener had to suffer lengthy neglect before his theory was proved correct.

Donald C. Johansen

Johansen, Donald C. (b. 1943) American anthropologist who found fossil human ancestors (hominids) at Hadar, in Ethiopia, that dated back 3 to 4 million years and were more primitive than *Australopithecus africanus*. One female skeleton fossil was nicknamed Lucy, after the Beatles' song. Johansen was the first to propose that some hominid body parts evolved to the human level before others.

Johanssen, Wilhelm (1857–1927) Danish botanist who introduced the term *gene,* and the terms *genotype* for the whole genome, and *phenotype* for the resulting characteristics of the organism. He showed that differences between genetically identical plants are solely of environmental origin.

Joly, John (1857–1933) Irish geologist who devoted himself to determining the age of the Earth (geochronology). A method using the salinity of the oceans proved inadequate and he took to radioactive decay as a source of information. He came up with an estimated age of 100 million years. The current estimate is about 4.6 billion years.

de Jussieu, Antoine Laurent (1748–1836) French taxonomist who demonstrated that the organizational elements in a plant are the best way to classify it. Most of his classifications are still used today. Jussieu believed that the most important difference between living and nonliving things is organization, which is possessed by the former but not the latter. He lived before the dawn of digital computers.

Johann Kepler

Kapteyn, Jacobus Cornelis (1851–1922) Dutch astronomer noted for his star catalog, which had more than a half million entries. He found new ways of determining stellar parallax, and constructed a model of our galaxy.

Kepler, Johann (1571–1630) German astronomer who also made discoveries in optics, general physics, and geometry. Kepler proved that the planets do not orbit in circles but in ellipses, with the Sun as the focus. He showed that the radius vector of each planet describes equal areas in equal time. Kepler described the supernova of 1604.

Knight, Andrew (1758–1838) English botanist who studied the patterns of growth of plant roots and stems. He provided some of the first descriptions of plant responses to external stimuli, a phenomenon known as tropism.

Kuiper, Gerard Peter (1905–73) Dutch-born American astronomer, involved with early American spaceflights, who, in 1948–49, discovered two new satellites: Miranda, the fifth satellite of Uranus, and Nereid, the second satellite of Neptune.

de Lacaille, Nicolas (1703–72) French astronomer and mathematician who charted many new constellations, naming them after scientific instruments. He was noted for establishing that the Earth is not a sphere but an oblate spheroid, having an equatorial diameter greater than its polar diameter.

Nicolas de Lacaille

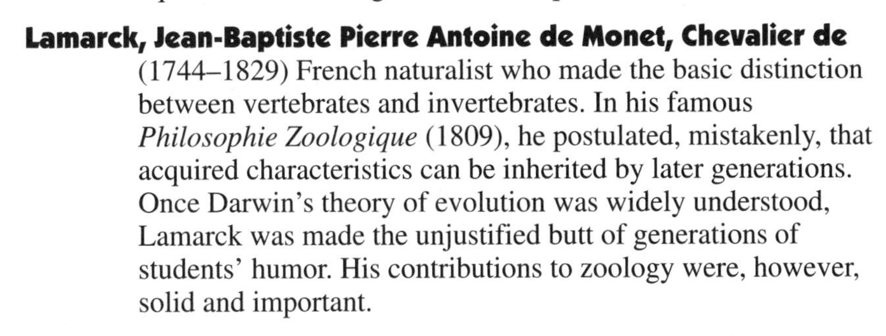

Lamarck, Jean-Baptiste Pierre Antoine de Monet, Chevalier de (1744–1829) French naturalist who made the basic distinction between vertebrates and invertebrates. In his famous *Philosophie Zoologique* (1809), he postulated, mistakenly, that acquired characteristics can be inherited by later generations. Once Darwin's theory of evolution was widely understood, Lamarck was made the unjustified butt of generations of students' humor. His contributions to zoology were, however, solid and important.

Jean Lamarck

Lamb, Hubert Horace (b. 1913) British climatologist who made a detailed record of British climatic changes for the previous several hundred years and attempted to suggest the reasons. This work is of special interest in the current context of possible global warming.

Lapworth, Charles (1842–1920) English geologist who in 1879 introduced the Ordovician system of geological strata as a separate entity. This finally settled a bitter argument between two geological pundits, one of whom believed the system to be Upper Cambrian while the other believed it to be Lower Silurian.

Lartet, Edouard Armand Isidore Hippolyte (1801–71) French pioneer paleontologist who was the first to find fossils of both apes and extinct animals in one place, thereby overthrowing the claim of Cuvier that no fossils of either apes or humans could exist.

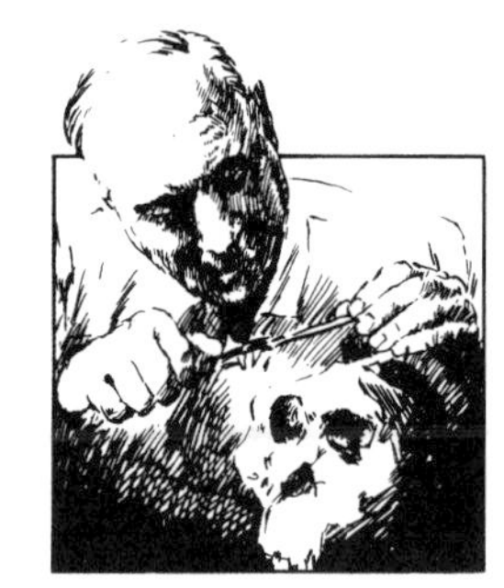

Louis Leakey

Leakey, Louis (1903–72) Kenya-born archaeologist and anthropologist who, with his wife, Mary Leakey, found hominid fossils in the Alduvai Gorge in East Africa, believed to be 1.75 million years old. These remains were later named *Australopithecus*. Leakey came to believe that this species became extinct and that the true ancestor of humans is *Homo habilis*, remains of which were found in 1960.

Leakey, Mary (1913–96) English archaeologist, anthropologist and wife of Louis Leakey, who in 1959 uncovered the skull of the hominid, later named *Australopithecus*, in the Alduvai Gorge in East Africa. In 1964, Mary Leakey unearthed fossilized trails of hominid footprints in volcanic ash, showing that hominids were walking upright 3.75 million years ago.

Leakey, Richard Erskine Frere (b. 1944) Kenyan paleontologist, son of Louis and Mary Leakey, who found in Kenya some of the oldest known homonid fossils, including a nearly complete fossil of a large *Homo erectus*. In 1989, he became director of the Wildlife Service in Kenya.

Leavitt, Henrietta Swan (1868–1921) American astronomer who showed that the apparent magnitude of brightness of Cepheid variable stars decreases linearly with the logarithm of the period of light variation. This work laid the basis for a method of measuring the distance of stars from Earth.

Lemaître, Abbe Georges (1894–1966) Belgian cosmologist and astronomer who in 1945 first proposed the "big bang" theory for the origins of the universe. He did not use this name, which was applied disparagingly by Fred Hoyle, but the term, and the theory, have achieved wide acceptance. The theory was also put forward by George Gamov about the same time.

Leucippus (fl. 500 BCE) Greek philosopher who is said to have originated the atomistic theory, which was taken up by Democritus and the poet Lucretius. The term *atom* derives from the Greek *a*, and *tomos*, "a cut," implying an entity so small that it cannot be subdivided further.

Leverrier, Urbain Jean Joseph (1811–77) French astronomer who, from observation of disturbances (perturbations) in the motions of planets, inferred the existence of an undiscovered planet and calculated the point in the heavens where, a few days later, Neptune was discovered by Galle (1846).

Carolus Linnaeus

Linnaeus, Carolus (Carl von Linné) (1707–78) Swedish naturalist and physician who introduced the binomial nomenclature of generic and specific names for animals and plants, which permitted hierarchical organization, later known as systematics. The generic name is always written with an initial capital letter, the specific name with a lowercase initial letter.

Sir Joseph Norman Lockyer

Lockyer, Sir Joseph Norman (1836–1920) English astronomer who wrote much on solar chemistry and physics, on meteorite hypotheses, and on the orientation of stone circles. He detected and named helium in the Sun's chromosphere in 1868.

Loomis, Elias (1811–89) American mathematician and astronomer who was a professor at the City University of New York (1844–60) and at Yale from 1860 to the year of his death.

Lorentz, Hendrik (1853–1928) Dutch physicist who, with George Fitzgerald in 1904, proposed a mathematical transformation, known as the Fitzgerald-Lorentz contraction, to account for the Mitchelson-Morley experiment's failure to show relative motion between the moving Earth and the postulated ether. The contraction is that of any moving body in the direction of its motion, and is negligible unless the speed approaches that of light. This idea helped to lead to Einstein's special theory of relativity.

Lorenz, Konrad Zacharias (1903–89) Austrian zoologist and student of animal behavior whose accounts of early imprinting, bonding, and aggression became very influential as a result of his highly successful popular books, among them *King Solomon's Ring* (1949) and *On Aggression* (1963). He is considered to be one of the founders of ethology.

Love, Augustus (1863–1940) English geophysicist who was the first to detect seismic waves transmitted over the surface of the Earth. These are called Love waves. His work contributed notably to the assessment of the differences in thickness of the Earth's crust below the continents and oceans.

Sir Bernard Lovell

Lovell, Sir [Alfred Charles] Bernard (b. 1913) English radio astronomer who pioneered the discipline. In the late 1940s and early 1950s, Lovell made observations on sporadic meteors, using radar, and succeeded in determining meteor trail heights and velocities. He is also remembered for having identified the first (Russian) artificial Earth satellite.

Percival Lowell

Lowell, Percival (1855–1916) American astronomer, best known for his mistaken conviction, prompted by Schiaparelli's use of the term *canali* ("channels"), that there were canals, built by intelligent beings, on Mars. Lowell wrote books to "prove" that Mars was, or had been, inhabited. His real scientific work was to predict, in 1905, the existence of a ninth planet beyond Neptune. This planet, named Pluto, was discovered by Tombaugh in 1930.

de Luc, Jean André (1727–1817) Swiss geologist and meteorologist who attempted to reconcile geology with the biblical account of the creation of the world. He suggested that the Flood was the result of a land collapse that caused the oceans to pour in, exposing the ocean floors, which then became continents. This provided an explanation for the puzzling fact that marine fossils are to be found in rock at the center of the continents.

Lucretius (fl. c. 100 BCE) Roman poet and philosopher who wrote a long poem called *De rerum natura* (On the nature of things) in which he outlined the atomist theory. He described a large army, seen from a remote cliff top, which appeared to be a solid body, but which reason confirmed to be composed of individual particles.

Sir Charles Lyell

Lyell, Sir Charles (1797–1875) Scottish geologist whose *Principles of Geology* (1830–33) had a powerful influence on modern scientific thought. In it, he denied the necessity of stupendous upheavals, arguing that the greatest geological changes might have been produced by forces that were still at work. The book greatly influenced Darwin. Lyell held that fossils are the best guides to identifying geological strata, and was one of the first to suggest that the Earth was much older than the 6,000 years suggested by the Bible.

de Maillet, Benoit (1656–1738) French naturalist who, while accepting that the biblical account of the Creation was true, insisted, nevertheless, that new animal and plant forms had come into existence during the history of the Earth.

Gideon A. Mantell

Mantell, Gideon A. (1790–1852) English geologist, who, with his wife, Mary, was the first to recognize that dinosaur bones are the remains of old extinct giant reptiles.

Marbut, Curtis Fletcher (1863–1935) American soil specialist (pedologist) who applied Russian ideas of soil formation to the United States. He was the first to produce a soil classification system in America.

Marsh, Othniel Charles (1831–99) American paleontologist who discovered (mainly in the Rocky Mountains) more than 1,000 species of extinct American vertebrates. Marsh searched for fossils, especially of dinosaurs, in the western United States, entering into fierce, and not always scrupulous, competition with other fossil hunters, especially Edward Cope.

Marsigli, Luigi (1658–1730) Italian naturalist who made measurements of the temperatures and pressures at varying depths in the oceans.

Maury, Matthew Fontaine (1806–73) American oceanographer whose detailed studies and analysis of ocean currents and winds, mainly collected from merchant seamen, enabled him to publish charts for navigators that substantially shortened ocean voyages for sailing ships. He made similar studies of ocean depths across the Atlantic.

Ernst Walter Mayr

Mayr, Ernst Walter (b.1904) German-born American zoologist whose early work was on the ornithology of the Pacific, but later was best known for neo-Darwinian views on evolution, as developed in *Animal Species and Evolution* (1963) and *Evolution and the Diversity of Life* (1976). Mayr also worked on the classification of organisms, on population genetics, and wrote *Population, Species, and Evolution* (1970).

Dmitri Ivanovich Mendeleyev

Mendeleyev, Dmitri Ivanovich (1834–1907) Russian chemist who organized the known elements into a table of columns arranged by chemical properties. This was the periodic table, which was of essential importance in the development of chemistry and enabled him to predict the existence of several elements discovered subsequently. Element 101 is named mendelevium after him. He also worked on gases, solutions, and petroleum.

Mercalli, Giuseppe (1850–1914) Italian seismologist who invented a scale of earthquake intensity based on the degree of damage caused by the ground movement. The Mercalli scale is not used much today, having been replaced by the Richter scale.

Mercator, Gerardus (Gerhard Kremer) (1512–94) Flemish geographer and map-maker who in 1568 introduced the map projection that bears his name and that, among others, has been used ever since. In 1585, he published a book of maps of Europe, completed by his son in 1595, which was the first to use the word *atlas*.

Messier, Charles (1730–1817) French astronomer who, while searching for comets, of which he discovered 13, mapped over 100 faint unmoving objects in the sky, which were actually nebulae (star clusters). He published the first nebula catalog in 1784. The prefix "M" is still applied to these objects.

Miller, Stanley (b. 1930) American chemist who studied the possible origins of life on Earth by using laboratory equipment to simulate the supposed early atmospheric gaseous content, and electric sparks to simulate lightning. He succeeded in forming amino acids, the units of proteins. Later work on the enzymatic function of RNA added credibility to Miller's ideas.

Miller, William Hallowes (1801–80) English mineralogist who classified crystals on the basis of their coordinates, and wrote one of the first treatises on crystallography.

Hugo von Mohl

Milne, John (1859–1913) English seismologist who became a professor of geology in Japan and a leading expert on the subject there. He was a pioneer of seismology and of the quantitative measurement of earthquakes. In 1880, he invented the modern seismograph, an instrument to record the horizontal movements of the Earth during earthquakes. He developed methods of locating distant earthquakes.

von Mohl, Hugo (1805–72) German botanist and pioneer of the microscopic study of plant structure and of research into plant physiology. He was the first to recognize protoplasm, now known as cytoplasm, as the principal substance of cells. He was also one of the earliest workers fully to understand and explain osmosis.

Mohorovičić, Andrija (1857–1936) Croatian geophysicist who during studies of earthquakes discovered the Moho discontinuity, which is the boundary between the Earth's crust and the mantle.

Mohs, Friedrich (1773–1839) German. mineralogist who wrote *The Natural History System of Mineralogy* (1821), and *Treatise on Mineralogy* (three volumes, 1825). He classified minerals on the basis of hardness, and the Mohs' scale of hardness is still in use. This scale, 0–10, is based on the ability of any mineral to scratch one lower down the scale. Talc is 1, diamond is 10.

Morley, Edward William (1838–1923) American chemist and physicist who, along with Albert Michelson, developed a sensitive interferometer, with which they showed that the speed of light is constant whether measured in the direction of the Earth's movement or perpendicular to that direction (Michelson-Morley experiment).

Sir Roderick Murchison

Murchison, Sir Roderick (1792–1871) Scottish geologist who first described various layers of ancient rock, including those of the Silurian and Devonian periods.

Naudin, Charles (1815–99) French botanist and horticulturist who experimented with plant hybridization, and found that certain characteristics are inherited on a regular basis.

Newlands, John Alexander Reina (1836–98) British chemist who, like Mendeleyev, was one of the first to show that the properties of chemical elements change in a periodic manner. He arranged the 62 elements then known in order of increasing atomic weight, and showed that these could be placed into groups of eight based on similar properties. This is known as the law of octaves.

Sir Isaac Newton

Newton, Sir Isaac (1642–1727) English scientist and mathematician, who became Lucasian professor of mathematics in 1669, a position now held by Stephen Hawking. By 1684, he had demonstrated the whole of his famous gravitation theory, supposedly inspired by the fall of an apple in his garden, in 1665 or 1666. Independently of Leibniz, he discovered the differential calculus. He also discovered that white light is composed of many colors, and invented a reflecting telescope. His great work, *Philosophiae naturalis principia mathematica* (1687), established him as the leading scientist of his day.

Noddack, Walter (1893–1960) German chemist who, working with his wife, Ida, discovered the elements rhenium and technetium, and did research on the photopigments of the eye.

Oldham, Richard D. (1858–1936) Irish geologist and seismologist who was the first to distinguish primary and secondary seismic waves, and who made many other contributions to geophysics.

Omalius d'Halloy, Jean-Baptiste-Julien (1783–1875) Belgian geologist who produced systematic subdivisions of geological formations and gave Cretaceous rocks their name.

Jan Handrik Oort

Oort, Jan Handrik (1900–92) Dutch astronomer who was the first to recognize that our galaxy is rotating. He also measured our distance from the center of the galaxy and produced an estimate of its total mass. He theorized that comets arose from a spherical shell of cometary ingredients surrounding the solar system.

Oppel, Albert (1831–65) German geologist and paleontologist who subdivided geological stages into zones. A zone fossil is a member of a group, such as ammonites or graptolites, that existed continuously over a particular span of geological time and can therefore be used to date the rock in which it is found.

d'Orbigny, Alcide Dessalines (1802–57) French founder of micropaleontology who categorized geological formations into stages.

Sir Richard Owen

Owen, Sir Richard (1804–92) English zoologist and comparative anatomist who opposed Darwin's evolution theory.

Paschen, Louis (1865–1947) German physicist who proved that the helium found on Earth is the same as that on the Sun. He is remembered by a series of lines in the hydrogen spectrum that were named after him.

Penck, Albrecht (1858–1945) German geographer and geologist who examined a sequence of past ice ages and provided a basis for later work on the European Pleistocene. He identified six topographic forms and reputedly introduced the term geomorphology. His classic work, *Morphology of the Earth's Surface*, was published in 1894.

Penck, Walther (1888–1923) German geologist and geomorphologist, the son of Albrecht Penck, who worked on theories of landform development, showing how a river cuts through a valley and how topological features are also caused by upthrusts of land.

Pfeffer, Wilhelm (1845–1920) German botanist who researched osmosis and the permeability of protoplasts (the content of the plant cell, including the cell membrane, within the cell wall) and concluded that the latter could be actively modified.

Wilhelm Pfeffer

Philolaus (fl. c. 480 BCE) Greek philosopher who suggested that the Earth was not the center of the universe, as was then and for centuries later generally believed, but that it moved through space. He taught that the Sun, Moon, and the known planets circled in separate spheres about a central fire, and that the Sun was a reflection of the latter.

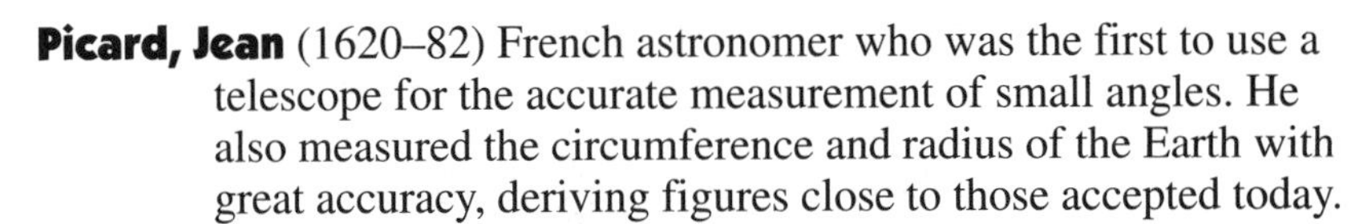

Picard, Jean (1620–82) French astronomer who was the first to use a telescope for the accurate measurement of small angles. He also measured the circumference and radius of the Earth with great accuracy, deriving figures close to those accepted today.

Edward Charles Pickering

Pickering, Edward Charles (1846–1919) American astronomer who specialized in the measurement of the brightness of stars (stellar photometry). Under his direction, a catalog of the magnitude of more than 45,000 stars was published, and an accurate map of the entire sky to magnitude 11 (1903) was produced. He compared the blue and the yellow brightness of stars (the color index), thereby being able to assess their temperature.

Playfair, John (1748–1819) Scottish mathematician, geologist, and philosopher who expanded on the work of his friend James Hutton. He was one of the main supporters of the doctrine of plutonism, which holds that the origin and nature of the whole of the Earth's crust are explicable by heat and that even sedimentary rocks acquired their final form by heat and pressure from below.

Pliny the Younger (Gaius Plinius Caecilius Secundus) (c. 61–112) Roman senator who is best remembered for his detailed account of the catastrophic volcanic eruption of Mount Vesuvius that destroyed Pompeii, southeast of Naples, Italy, in 79 CE.

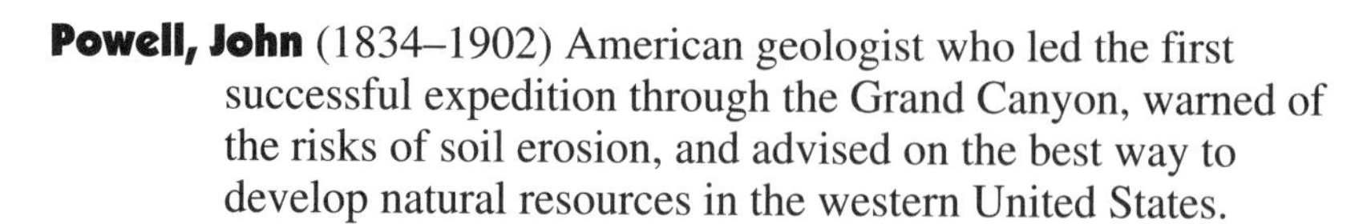

Powell, John (1834–1902) American geologist who led the first successful expedition through the Grand Canyon, warned of the risks of soil erosion, and advised on the best way to develop natural resources in the western United States.

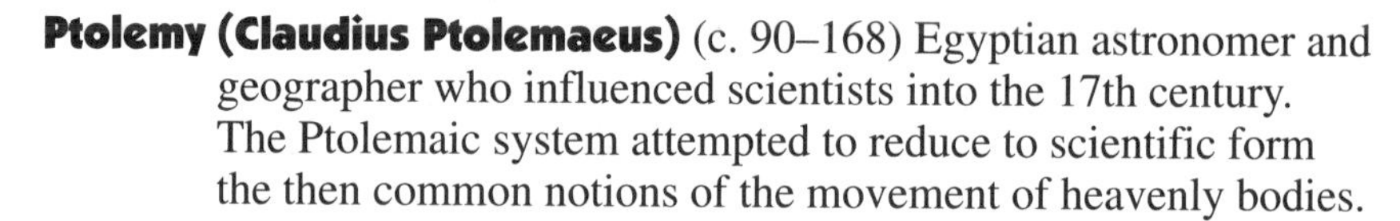

Ptolemy (Claudius Ptolemaeus) (c. 90–168) Egyptian astronomer and geographer who influenced scientists into the 17th century. The Ptolemaic system attempted to reduce to scientific form the then common notions of the movement of heavenly bodies.

Ptolemy

Pytheas (c. 350 BCE) Greek geographer and explorer who studied the strong Atlantic tides and concluded correctly that they were somehow caused by the Moon.

Sir William Ramsay

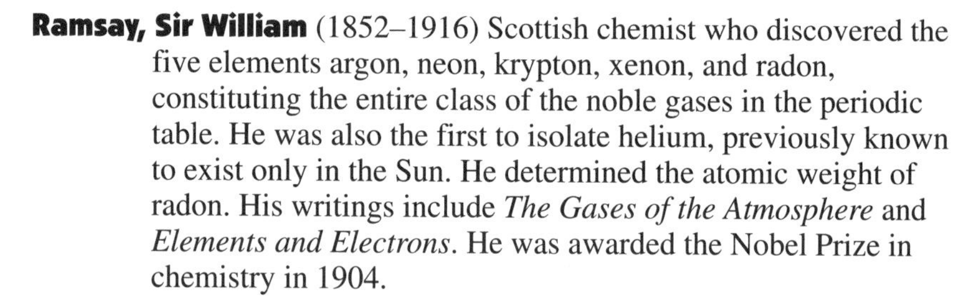

Ramsay, Sir William (1852–1916) Scottish chemist who discovered the five elements argon, neon, krypton, xenon, and radon, constituting the entire class of the noble gases in the periodic table. He was also the first to isolate helium, previously known to exist only in the Sun. He determined the atomic weight of radon. His writings include *The Gases of the Atmosphere* and *Elements and Electrons*. He was awarded the Nobel Prize in chemistry in 1904.

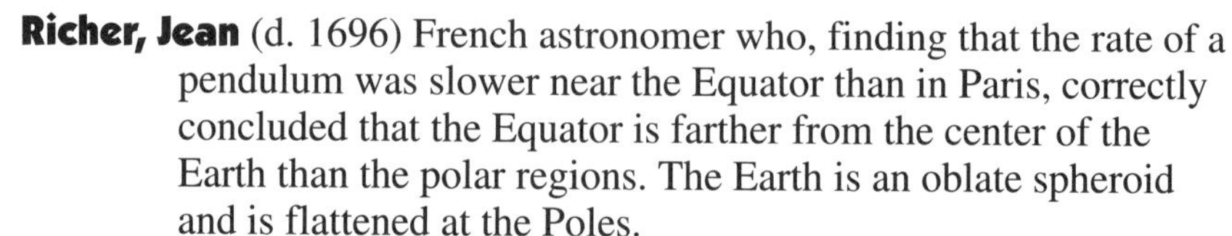

Richer, Jean (d. 1696) French astronomer who, finding that the rate of a pendulum was slower near the Equator than in Paris, correctly concluded that the Equator is farther from the center of the Earth than the polar regions. The Earth is an oblate spheroid and is flattened at the Poles.

Richter, Charles Francis (1900–85) American seismologist who, with Gutenberg, devised an absolute scale of earthquake strength, based on the logarithm of the maximum amplitude of earthquake waves on the seismograph, adjusted for distance from the earthquake epicenter. The scale bears his name.

Roemer, Olaus (1644–1710) Danish astronomer who noticed that the time between eclipses of a satellite of Jupiter was greater when Jupiter and the Earth were moving away from each other and less when they were approaching each other. He concluded that light must be taking longer to reach Earth from the greater distance and that it must therefore move with a finite speed. He was able to make an assessment of the speed.

Russell, Henry Norris (1877–1957) American astronomer and professor of astronomy at Princeton University who suggested a now superseded theory of stellar evolution in which stars begin as dim dull objects, shrink and heat up, and then cool again to become red stars once more.

Carl Edward Sagan

Rutherford, Daniel (1749–1819) Scottish chemist who was one of the discoverers of nitrogen gas.

Sagan, Carl Edward (1934–96) American astronomer who, through books and television, did much to popularize this aspect of science. He worked on the physics and chemistry of planetary atmospheres and surfaces, and on the origin of life on Earth.

Matthias Jakob Schleiden

de Saussure, Horace Bénédict (1740–99) Swiss geologist and physicist who explained the shape of the Alps by horizontal opposing movements of strata causing levels to be pushed up (alpine folding), and by subsequent erosion. He wrote extensively on mineralogy, botany, and meteorology.

Schiaparelli, Giovanni Virginio (1835–1910) Italian astronomer and head of the Brera observatory in Milan who studied meteors and twin stars and, in 1877, described the so-called canals of Mars, and in 1861, the asteroid Hesperia.

Schleiden, Matthias Jakob (1804–81) German botanist who did much to establish cell theory. He showed that cells are the units of structure in plants and animals, and that organisms are aggregates of cells arranged according to definite laws.

von Schlotheim, Ernst (1764–1832) German paleontologist and pioneer who used fossils to find the relative ages of rock layers.

Seaborg, Glenn T. (1912–99) American chemist and atomic scientist who discovered many previously unknown isotopes of common elements. He assisted in the production of a number of non-natural, above-uranium (transuranic) elements, including neptunium (93), plutonium (94), americium (95), berkelium (97), einstinium (99), fermium (100), and nobelium (102). During his lifetime, he was honored by the naming of element 106 as seaborgium. He was also involved in the production of the fissionable isotope plutonium-239, which has formed the basis of atomic weapons ever since. He shared the 1951 Nobel Prize in chemistry.

Sedgwick, Adam (1785–1873) English geologist who carried out geological mapping in Wales and introduced the term *Cambrian* for rocks formed in the first 100 million years of the Palaeozoic era, during which time marine invertebrates, especially trilobites, flourished. He also named the Devonian period after geological studies in southwest England.

Sefström, Nils G. (1765–1829) Swedish chemist who in 1880 discovered vanadium, the metallic element later alloyed with steel to produce very high-strength, low-corrosion metal for tools and other purposes. Vanadium was actually discovered by Andres del Rio in 1801, but he let himself be persuaded that the substance he had found was an impure form of chromium.

Segrè, Emilio (1905–89) Italian-born American physicist who discovered the non-natural element technetium and helped to produce plutonium. His research team discovered the antiproton.

William Smith

Shapley, Harlow (1885–1972) American astronomer who proved that our solar system is near the edge of the galaxy rather than being central, as had been supposed.

Smith, William (1769–1839) English civil engineer, known as the founder of English geology, who used fossils to identify sedimentary rock layers and produced a massive *Geological Map of England* (1815), as well as 21 color-coded geological maps of English counties (1819–24).

Steller, Georg Wilhelm (1709–46) German naturalist and explorer who studied numerous species, including the sea lion, the eider duck, and the jay that was named after him, the Steller jay.

Stensen, Niels (Nicolaus Steno) (1638–86) Danish physician, naturalist, and theologian who did fundamental work in anatomy, geology, crystallography, paleontology, and mineralogy. He was the first to point out the true origin of fossil animals (1669), to explain the structure of the Earth's crust, and to distinguish between sedimentary and volcanic rocks.

Struve, Friedrich Georg Wilhelm (1793–1864) German-born Russian astronomer whose studies of binary stars showed that Newton's law of gravitation operates outside the solar system and is, therefore, presumably, universal.

Friedrich Georg Wilhelm Struve

Suess, Eduard (1831–1914) Austrian geologist, founder of "new geology." His theory that there had once been a great supercontinent made up of the present southern continents led to modern theories of continental drift.

Eduard Suess

Tansley, Sir Arthur George (1871–1955) English botanist who pioneered the science of plant ecology. He was also active in promoting various organizations devoted to wildlife preservation and ecology, such as the British Ecological Society and the Nature Conservancy.

Thales (c. 624–545 BCE) Greek astronomer, mathematician, and physicist who is said to have correctly predicted a solar eclipse, and who believed that everything was compounded from water.

Theophrastus (c. 372–c. 287 BCE) Greek philosopher and botanist most of whose writings have been lost, but who, in *Historia plantarum* and *Plantarum causae*, recorded some 450 plant species. He described features of plant organization, distinguishing these from those of animals, and the medical uses of plants. Much later botany was based on these works.

Thomson, Sir Charles Wyville (1830–82) British marine biologist who showed that living creatures exist at great depths in the oceans. He directed the round-the-world scientific voyage of HMS *Challenger* (1872–76), and was also a pioneer of the science that became modern oceanography.

Thunberg, Carl P. (1743–1828) Swedish botanist who amassed one of the largest collections of botanical specimens of his time. He was also the first Western botanist to interest himself in Japanese flora and wrote extensively on them in *Flora Japonica* (1784). He also studied the flora of South Africa.

Tinbergen, Nikolaas (1907–88) Dutch-born British zoologist and ethologist who made a major study of animal behavior in the wild. He and Konrad Lorenz are considered to be the founders of ethology. Tinbergen shared the 1973 Nobel Prize in physiology or medicine with Lorenz and Karl von Frisch.

Tombaugh, Clyde William (1906–97) American astronomer and professor at California University, where, in 1930, he discovered the planet Pluto by looking for differences between two photographs taken a little time apart.

Tradescant, John (1570–1633) British naturalist, botanist, traveler, and head gardener to King Charles I, who was a notable pioneer in the collection and cultivation of plants, and who introduced many new species of plants into English gardens. He was the first to open a museum to the public, in Lambeth, London.

Tradescant, John (1608–62) British botanist and plant collector, son of John Tradescant, who continued his father's work as a collector and cultivator of plants. He collected specimens in Virginia, and took over his father's job as head gardener to King Charles I. In his will, he left the museum he had inherited to Elias Ashmole, and this became the celebrated Ashmolean Museum in Oxford, England.

von Tschermak-Seysenegg, Erich (1871–1962) Austrian botanist who drew attention to the long-neglected work of Gregor Mendel, thereby arousing new interest in genetics.

Tswett or **Tsvett, Mikhail Semenovich** (1872–1919) Russian botanist who in 1906 devised a percolation method of separating plant pigments, thus making the first chromatographic analysis.

Jethro Tull

Tull, Jethro (1674–1741) English lawyer, farmer, writer, and inventor who devised a machine for planting seeds in rows. He had a major influence on plant cultivation, suggesting the use of manure and the importance of hoeing around plants to remove weeds so as to reduce competition for nutrients.

Urbain, Georges (1872–1938) French chemist who specialized in the study of the rare earth elements. After enormous labor involving hundreds of thousands of fractional crystallizations, he discovered samarium, europium, gadolinium, terbium, dysprosium, holmium, lutetium, and hafnium.

Van Allen, James Alfred (b. 1914) American physicist who developed the radio proximity fuse; employed rockets to study the physics of the upper atmosphere; contributed to the success of the first American artificial satellite (*Explorer 1*); and discovered two belts of energetically charged particles circling the Earth (Van Allen radiation belts), which are retained by the Earth's magnetic field.

de Vries, Hugo (Marie) (1848–1935) Dutch botanist and geneticist who from 1890 devoted himself to the study of heredity and variation in plants, significantly developing Mendelian genetics and evolutionary theory. *Die Mutationstheorie* (The mutation theory), in which he showed that mutations occur in organisms, was published 1901–3.

Hugo de Vries

Wallace, Alfred Russel (1823–1913) Welsh naturalist whose memoir, sent to Darwin in 1858 from the Moluccas in the East Indies, was read at a meeting of the Linnaean Society at which Darwin's paper was also read. Wallace's paper virtually duplicated Darwin's theory of evolution by means of natural selection, and hastened Darwin's publication of *On the Origin of Species*. This revolutionary work was extended by Wallace's *Contributions to the Theory of Natural Selection* (1870). Wallace was generous and apparently free from jealousy. He called his own later book on evolution *Darwinism* (1889). He recorded the division between zoological types in the east and west islands of Malaysia, known as the Wallace line.

Alfred Lothar Wegener

Wegener, Alfred Lothar (1880–1930) German geologist and meteorologist who, having noted that the Atlantic east coast could be fitted roughly into the Atlantic west coast, first suggested the idea of continental drift in 1912. His theory was detailed in *Origins of Continents and Oceans* (1915). He was laughed at, but by the 1960s measurements had proved that the continents had moved. His ideas became accepted and have helped to pave the way for the theory of plate tectonics, which has been established as one of the major tenets of modern geophysics.

Went, Friedrich August Ferdinand Christian (1863–1935) Dutch botanist and specialist in tropical agriculture, whose Utrecht School was renowned for its research into plant physiology.

Abraham Gottlob Werner

William Whewell

Werner, Abraham Gottlob (1750–1817) German geologist, one of the first to attempt classification of rocks. He popularized Neptunism, an erroneous theory claiming that almost all rocks were sedimentary and had been created from the water of an early universal ocean.

Whewell, William (1794–1866) English naturalist and philosopher of science who wrote about catastrophism, a term he invented for the belief that during the history of the Earth, one or more major convulsions must have occurred to account for the irregular shape of mountains, coastlines, gorges, and so on. Prior to the 19th century most geologists were catastrophists, but evidence that the Earth was much older than the biblical 6,000 years led to the abandonment of this unnecessary idea.

White, Gilbert (1720–93) British naturalist and clergyman, remembered for his best-selling book *The Natural History of Selborne* (1789), on the animal and plant life of Selborne in Hampshire, England, which has remained in print continuously for more than 200 years.

Xenophanes (fl. 600 BCE) Greek philosopher who explained fossils by proposing that the Earth had undergone successive floods. He criticized religions, pointing out that the gods were invariably conceived in human form (anthropomorphism) and so were simply human concepts.

Yonge, Charles Maurice (1899–1986) English marine biologist who studied coral physiology, oyster physiology, and the ecology of the Great Barrier Reef. He wrote a number of popular books on marine biology.

SECTION THREE CHRONOLOGY

c. 1760 BCE ● Babylonians compile detailed observations of movements of the planet Venus

c. 1750 BCE ● Mathematical and geometric knowledge recorded in Egyptian papyrus known as Moscow Papyrus

c. 950 BCE ● Natural gas from wells in use Chinese

720 BCE ● Solar eclipses recorded in China

700 BCE ● Signs of the zodiac identified by Babylonians

673 BCE ● Babylonian astronomers correctly predict solar eclipses

c. 570 BCE ● Early geological theory proposed by Greek philosopher Xenophanes on basis of fossil sea shells found on mountains miles from the sea

c.530 BCE
Pythagoras.

c. 530 BCE ● Concept of evolution proposed by Greek astronomer and philosopher Anaximander. Pythagoras, Greek philosopher and mathematician, argues that Earth is a sphere

480–471 BCE ● Greek philosopher Oenopides calculates Earth is tipped by 24 degrees with respect to the plane of its orbit

450–441 BCE ● Greek Pythagorean philosopher Philolaus suggests there is a central fire around which the Earth moves and that the Earth rotates

c. 445 BCE ● Greek philosopher and mathematician Anaxagoras dies; he had been prosecuted for impiety for suggesting that the Sun is a hot stone rather than a deity, and that the Moon borrows light from the Sun

400 BCE ● Earliest known theory of a moving Earth is put forward by Philolaus, a member of the Pythagorean school

390–381 BCE ● Greek philosopher Plato suggests existence of a continent directly opposite Europe on the other side of the globe, which he calls the Antipodes

350 BCE ● Aristotle theorizes that the universe is arranged in concentric shells, with the Earth dominating the center. Star catalog is prepared by Chinese astronomer Shin Shen

314 BCE • First known geology text is written by Theophrastus

c. 250 BCE • Accurate estimate of Earth's circumference is made by Greek philosopher Eratosthenes

240–231 BCE • Greek philosopher Eratosthenes calculates that the circumference of the Earth is 28,500 miles (46,000 km)

164 BCE • Detailed observations are made of Halley's comet

c. 150 BCE • System of defining geographical locations by lines of longitude and latitude is established by Greek astronomer Hipparchus

c. 130 BCE • Distance and size of the Moon is calculated by Greek astronomer Hipparchus

130s BCE • Greek astronomer Hipparchus argues that the Earth is motionless at the center of the universe

c. 10 BCE • Ways of drilling wells to a depth of 4,800 feet (1,400 m) is devised in China

c. 1 BCE • World is divided into frigid, temperate, and torrid zones by Greek geographer Strabo

c. 20 CE • All existing geographical knowledge is compiled in 17 works, *Geographica*, of Greek geographer Strabo

79 • Earliest scientific account of volcanic eruption is written by Roman writer Pliny the Younger

c. 120 • System of locating points on a map with a grid is devised by Chinese geographer Zhang Heng

132 • Seismograph for measuring earthquake movements is built in China

c. 300 • Coal in use for making cast iron by Chinese

850 • Astrolabe is refined by Arab scientists

940 • Invention of a map-projection technique by Chinese astronomers, later called Mercator projection

c. 1000 • Coal in use as a fuel by Chinese

c.300
Chinese use coal for making cast iron.

940
Chinese Tunhuang star map.

1086 • Geological concepts of erosion, uplift, and sedimentation formulated in Chinese book

c. 1100 • Cause of solar eclipses is demonstrated by Chinese

1120 • Anglo-Saxon Welcher of Malvern introduces latitude and longitude measurements, in degrees, minutes, and seconds

c. 1190 • Magnetic compasses in use in Europe

1233 • Coal mining begins in Newcastle, England

1269 • First scientific account of magnetic poles and use of the compass

1298 • Coal and asbestos first described in Europe by Venetian explorer Marco Polo

c. 1350 • Concept of impetus developed by French philosopher Jean Buridan to explain motion of heavenly bodies

1430s • New star map published by Mongol astronomer Ulugh Beg

1514 • First version of a Sun-centered universe is written by Polish astronomer Nicolaus Copernicus (not published until 1543 because of fear of church persecution)

1517 • Fossils are explained as organic remains by Italian scholar Girolamo Fracastoro

1533 • Principle that longitude can be found by comparing time on a clock with the position of the Sun is discovered by Dutch geographer Reiner Dokkum

1543 • Copernicus' account of planetary movement based on a Sun-centered Universe, *Revolutionibus*, is published; theory will trigger a scientific revolution. Aristotle's theories of motion and space attacked by French scientist Petrus Ramus

1544 • Earliest compendium of world geography published by German theologian Sebastian Münster

1546 • Term *fossils* coined by German mineralogist Georgius Agricola for anything dug from the earth

1556 • German metallurist Georgius Agricola publishes detailed discussion of mineral vein locations

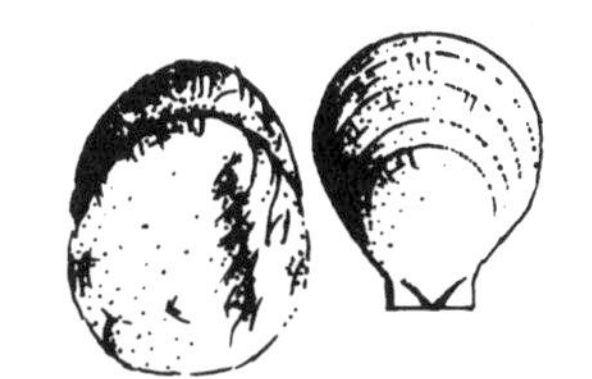

1546
Fossil ammonites.

1556
Drawing of a mine from Georgius Agricola's De Re Metallica.

1568 • Flemish cartographer Gerardus Mercator designs the first map to use Mercator projection

1570 • First comprehensive atlas of the world is published by Flemish geographer Abraham Ortelius

1577 • Comets are proved not to be atmospheric events by Danish astronomer Tycho Brahe

1583 • Principle of the pendulum is discovered by Italian scientist Galileo Galilei

1590 • Aristotle's theories of motion are refuted by Italian scientist Galileo Galilei

1597 • Copernican system of the universe is defended by Italian scientist Galileo Galilei in a letter to German scientist Johannes Kepler

c. 1600 • Theory that the Earth is a huge magnet is put forward by English physician William Gilbert, who also studies static electricity

1608 • Refracting telescope invented by Dutch instrument maker Hans Lippershey

1610 • Discovery of rings of Saturn, phases of Venus, moons of Jupiter, and irregular surface of the Moon by Italian scientist Galileo Galilei, using a telescope

1620 • English philosopher Francis Bacon observes the near fit of the west coast of Africa and the east coast of South America

1635 • Evidence that the Earth's magnetic poles shift position over time presented by English astronomer Henry Gellibrand

1643 • Barometer (instrument for measuring air pressure) invented by Italian physicist Evangelista Torricelli

1647 • French scientist Blaise Pascal demonstrates that air pressure decreases with altitude and shows that air has a finite height

1665 • English physicist Isaac Newton conceives the idea of universal gravitation and an explanation for lunar motions

1668 • Reflecting telescope invented by English physicist Isaac Newton who uses it to study celestial phenomena

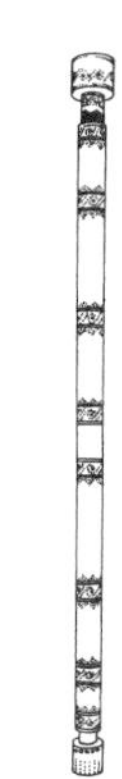

1610
Galileo's telescope.

1643
Torricelli invents barometer.

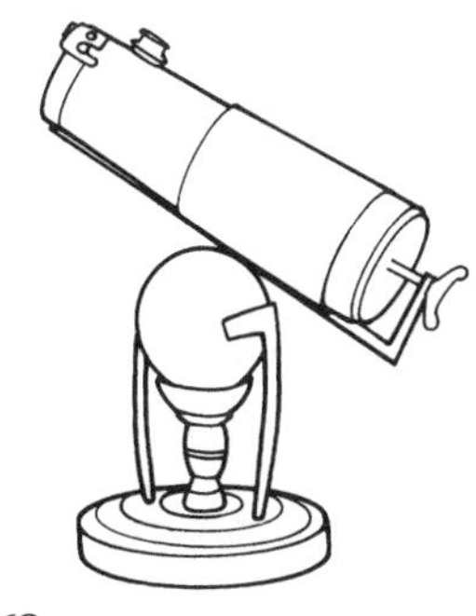

1668
Isaac Newton invents reflecting telescope.

1670s • French astronomer Jean Richer concludes that the diameter of the Earth is greater around the Equator than from Pole to Pole and that the Earth is not a perfect sphere

1674 • Study of the hydrologic cycle probably started by French lawyer Pierre Perrault, who solves the mystery of the origins of springs

1684 • First fairly accurate figure for the circumference and diameter of the Earth published posthumously in a book by French astronomer Jean Picard

1686 • Correct account of planetary orbits published by English physicist Isaac Newton. First meteorological world map produced by English mathematician and astronomer Edmund Halley

1687 • English physicist Isaac Newton theorizes that the Earth is an oblate spheroid. Newton's three laws of motion and law of universal gravitation published in *Philosophiae naturalis principia mathematica*

1698 • First scientific ocean voyage is undertaken by English mathematician Edmund Halley in order to map magnetic declinations of the globe

1702 • First astronomical textbook based on Newtonian principles published by Scottish mathematician David Gregory

1714 • Mercury thermometer invented by German physicist Gabriel Fahrenheit

1715 • In Europe astronomers gather to observe a total solar eclipse

1721 • Frenchman Henri Gautier accounts for the process of river erosion

1725 • *Historia Coelestis Britannica*, a work listing 3,000 stars and compiled under the direction Britain's astronomer royal John Flamsteed, is published posthumously

1735 • Confirmation of Isaac Newton's theory that the Earth is an oblate spheroid and not a perfect sphere. George Hadley models the Earth's wind circulation

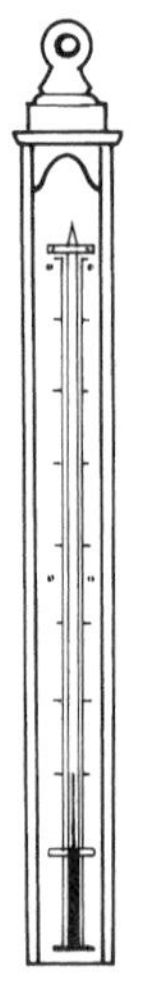

1714
Gabriel Fahrenheit's thermometer.

1738 • Dutch diplomat Benoit de Maillet, argues that the Earth's surface was shaped by a universal ocean

1743 • Early geological map drawn by British naturalist Christopher Packe

1749 • French naturalist Georges Louis Leclerc, Comte de Buffon, speculates that the Earth was formed 75,000 years ago after a comet collided with the Sun [or 1745?]

1750 • First catalog of Southern stars is compiled by French astronomer Nicolas de Lacaille

1755 • Increased interest in geological phenomena when Lisbon in Portugal suffers a disastrous earthquake

1757 • Sextant invented by Scottish inventor John Campbell

1761 • Correct determination of longitude using English horologist John Harrison's chronometer on a voyage to Jamaica. Publication of an article dealing with stratigraphy by German physician George Christian Fuschel

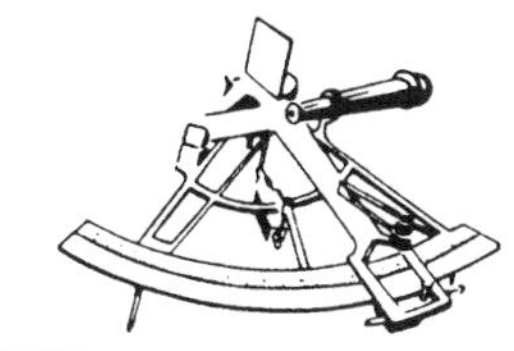

1757
Sextant invented.

1765 • Prismatic basalt discovered to be of volcanic origin by French geologist Nicolas Desmarest

1775 • Improved anemometer (instrument for measuring wind speed) invented by Irish physicist James Lind

1779 • Term *geology* coined by Swiss geologist Horace Bénédict de Saussure

1784 • A balloon is flown over London by American physician John Jeffries in order to collect air samples at various heights

1797 • High-pressure, high-temperature mineralogy is pioneered by Scottish geologist Sir James Hall, who also proves igneous rock cools to form crystalline rock

1798 • Catalog of stars published by German astronomer Caroline Herschel

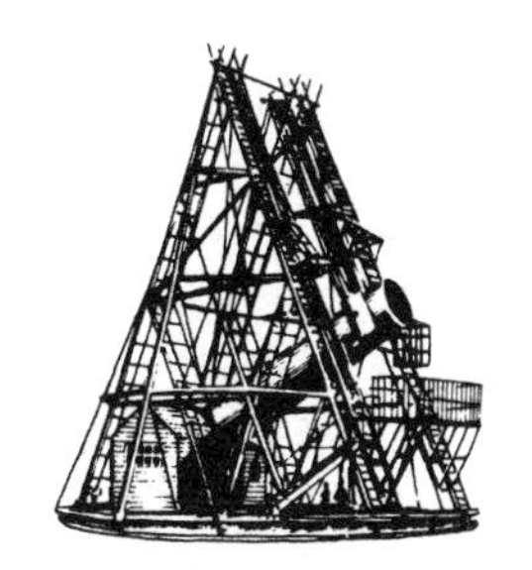

1798
Caroline Herschel publishes star catalog.

1799 • Jurassic period in the Earth's history is identified by German Alexander von Humboldt. William Smith, English geologist,

proposes that rock strata can be identified by their characteristic fossils

1804
Spherical baloon used for observation and research from 1783.

1804 • French scientists Jean-Baptiste Biot and Joseph-Louis Gay-Lussac ascend in a balloon to study the atmosphere and the Earth's magnetic field at high altitudes

1806 • Wind scale is devised by British naval officer Francis Beaufort

1807 • Founding of the Geological Society of London

1815 • Publication of rock classification based on fossil remains by English geologist William Smith

1819 • Founding of the American Geological Society

1822 • Cretaceous period in the Earth's history is identified by Jean-Baptiste-Julien d'Halloy. First fossil dinosaur, an iguanodon, discovered by Mary Mantell. German mineralogist Friedrich Mohs introduces a system for classifying minerals and a scale for mineral hardness. Publication of a textbook on stratigraphy by William Daniel Conybeare and William Phillips, who also identify the Carboniferous period

1829 • French geologist Alexandre Brongniart names the Jurassic period in Earth's history

c. 1830 • Development of doctrines on the origin of mountain ranges by French geologist Élie de Beaumont

1830 • Principle of uniformitarianism (that the Earth's surface features can be interpreted as the result of action by physical, chemical, and biological processes through time) established by Scottish geologist Charles Lyell

1831 • Chart of the structure of a hurricane is published by American meteorologist William Redfield

1833 • Completion of a star catalog by German astronomer Friedrich Wilhelm Bessel that contains 50,000 stars. Scottish geologist Charles Lyell identifies Recent, Pliocene, Miocene, and Eocene periods in the Earth's history. Stroboscope invented by Austrian scientist Simon von Stampfer

1834 • Triassic period in the Earth's history identified by Friedrich August von Alberti. Understanding of glaciers is advanced by geologist Swiss geologist Johann von Charpentier

1835 • Silurian period in the Earth's history identified by Scottish geologist Roderick I. Murchison. Cambrian period in the Earth's history named by English geologist Adam Sedgwick

1839 • Devonian period in the Earth's history named by English geologist Adam Sedgwick and Scottish geologist Roderick I. Murchison. First official observatory in the United States is founded at Harvard College

1840 • Theory of ice ages is advanced by Swiss zoologist and geologist Louis Agassiz

1841 • Arnold Escher von der Linth describes mountain folds in layers of rock, later known as "nappes." Permian period in Earth's history identified by Scottish geologist Roderick I. Murchison

1846 • Publication of the first synoptic weather map by American professor Elias Loomis

c. 1850 • Term *seismology* coined by Irish engineer Robert Mallet

1851 • Earth's rotation is proved by means of a pendulum built by French physicist Léon Foucault

1852 • Gyroscope constructed by French physicist Léon Foucault

1855 • First textbook on oceanography is published by American oceanographer Matthew Maury

1856 • Remains of early human ancestor are found in the Neander Valley, Germany, and named Neanderthal Man

1858 • Theory of continental drift proposed by American geologist Antonio Snider-Pellegrini

c. 1862 • Theories of uniformitarianism and Darwinian evolution are attacked by Scottish physicist William Thomson (later Lord Kelvin)

1863 • Modern weather-mapping techniques introduced by English scientist Francis Galton, who also produces theory of

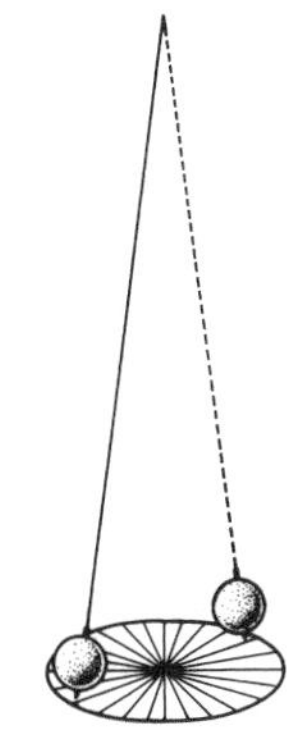

1851
Foucault's pendulum constructed.

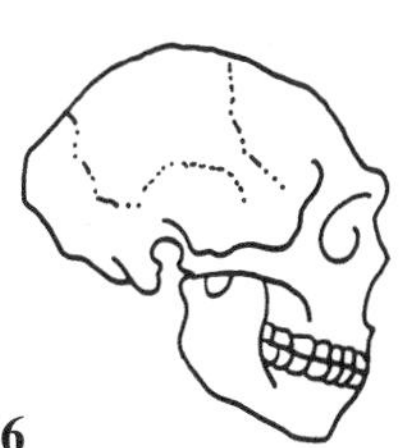

1856
Neanderthal Man discovered.

1863
Francis Galton's Synchronous Weather Chart.

anticyclones. Greenhouse effect discovered by Irish physicist John Tyndall

1866 • Textbook on petrology published by German mineralogist Ferdinand Zirkel. Proposition that Earth has a core comprising iron and nickel put forward by French geologist Gabriel Auguste Daubrée

1873 • Independent treaties on microscopical petrography published by German mineralogist Ferdinand Zirkel and German geologist Karl Rosenbusch

1875 • Russian V. V. Dokudaev starts work from which will evolve the basic principles of pedology

c. 1876 • Norwegian meteorologist Henrik Mohn and mathematician Cato Maximilian Guldberg begin work in the field of dynamical meteorology

1879 • Invention of the meridian photometer by American astronomer E. C. Pickering

1880 • English geologist John Milne invents the seismograph

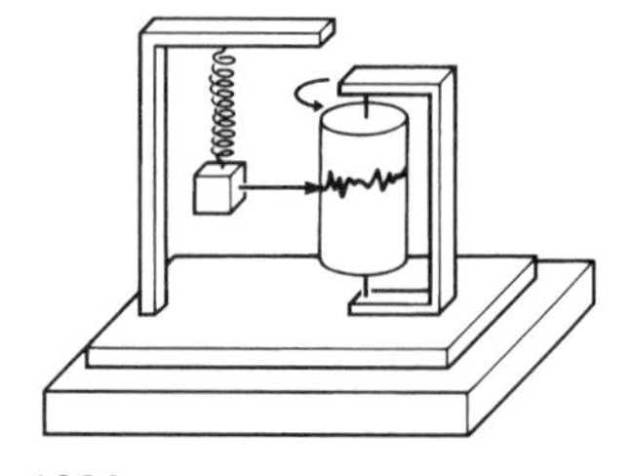

1880
Invention of the seismograph.

1884 • Prime meridian through Greenwich, England, is established. Vladimir Köppen produces world map of temperature zones

1885 • Englishman Lord Rayleigh identifies type of earthquake waves, now called Rayleigh waves

1889 • New method of landscape analysis introduced by American geologist Wilham Morris Davis

1892 • English physicist Oliver Heaviside discovers the ionosphere. Term *isostasy* introduced by American geologist Clarence Edward Dutton. First textbook on limnology (the study of lakes) presented by Swiss naturalist F. A. Forel

1896 • Concept of global warming formulated by Swedish chemist Svante Arrhenius

1897 • British geologist R. D. Oldham discovers that seismic waves consist of two components. Norwegian-American Jacob Bjerknes, with Vilhelm Bjerknes, helps develop the mathematical theory of weather forecasting

1902 • Troposphere and stratosphere are identified by French meteorologist Léon-Philippe Teisserenc de Bort

1904 • Publication of a study by Norwegian-American meteorologist Jacob Bjerknes that takes a scientific approach to weather forecasting

1910 • Modern tide theories developed by French mathematician Jules-Henri Poincaré

1911 • Augustus Love's *General Theory of Earth Waves* published, including descriptions of earthquake waves named after him

1912 • Publication of a work concerning the hydrology of the Mediterranean Sea by N. Nielson. Ultrasonic techniques used by Russian K. Silovski to detect icebergs and submerged ice. Theory of continental drift is proposed by German geologist Alfred L. Wegener

1913 • Large numbers of bottles are launched to study surface currents in the Sea of Japan. Existence of ozone layer is proved by French physicist Charles Fabry

c. 1920 • Atmosphere shown to be made up of air masses differing in temperature marked by sharp boundaries called fronts by Norwegian meteorologists Vilhelm and Jacob Bjerknes

1920s • First geochemical classification of elements proposed by Russian scientists. Development of petroleum geology

1926 • Galaxies classified as elliptical, spiral, or irregular by American astronomer Edwin Powell Hubble

1927 • "Big bang" theory of the universe's origin is proposed by Belgian astronomer Georges Lemaître, although the theory is not yet known by that name. German meteorologist Rudolf Geiger founds the study of microclimatology

1928 • Development of the first ultrasound echograph

1929 • Formation of the concept of geochemical migration of elements by Alexander Yevgenevic Fersman

1930s • "Big bang" theory popularized by Russian-American physicist George Gamov

1912
Wegener proposes his theory of continental drift.

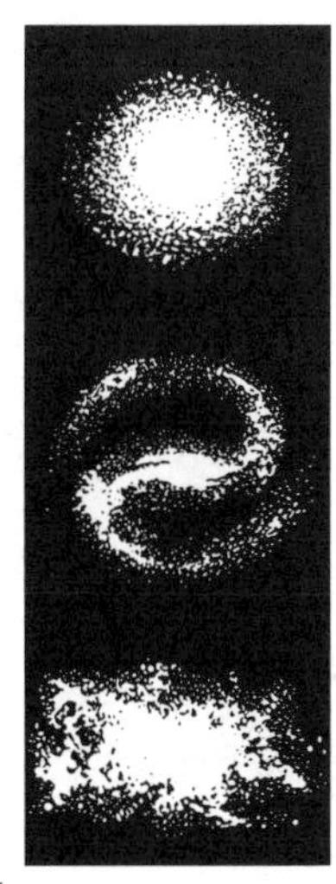

1926
Galaxy classification.

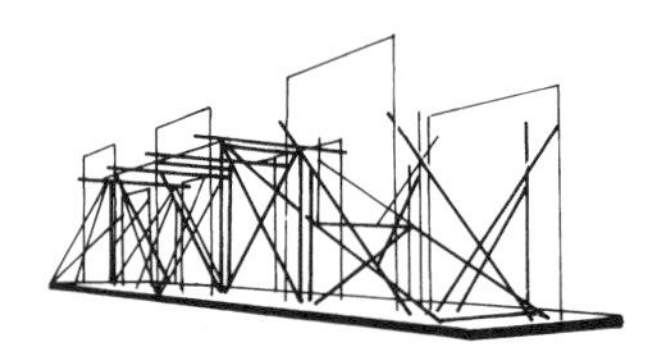

1931
Extraterrestrial radio waves discovered.

1931 • Karl Jansky discovers extraterrestrial radio waves with a rotatable aerial array

1931 • Radio astronomy founded by American Karl G. Jansky when he discovers radio radiation from outside the solar system

1935 • American seismologist Charles Richter develops the Richter scale for measuring intensity of earthquakes

1940 • Publication of seismic-wave travel timetables, known as the Jeffreys and Bullen tables

1940s • Discovery of the jet stream, a narrow, eastward wind current above the lower troposphere, by Norwegian-American meteorologist Jacob Bjerknes

1947 • Around this time, scientists abandon the idea that ocean floors are flat. Method of carbon dating discovered by American chemist Willard F. Libby

1950 • Adoption of a new astronomical unit, the ephemeridical unit, used to measure time

1950s • This decade marks the start of significant studies of continental shift, sea-floor spreading, and plate tectonics

1951 • Computer used for the first time to calculate planetary orbits

1954 • Invention of the microprobe for use in experimental mineralogy

1957 • International Geophysical Year; 70 countries engage in a coordinated study of the Earth and its atmosphere

1958 • Solar wind (a stream of charged particles emmited by the Sun) discovered by American physicist Eugene Parker

1959 • First television pictures of the Earth's cloud cover from space are taken by US satellite *Explorer 6*. Fossil remains of an early hominid 1.75 million years old are found in Kenya by Anglo-Kenyan anthropologist Mary Leakey

1960 • First use of geothermal power in the United States. First weather satellite launched. In the field of plate tectonics American geologist Henry H. Hess proposes the concept of seafloor spreading

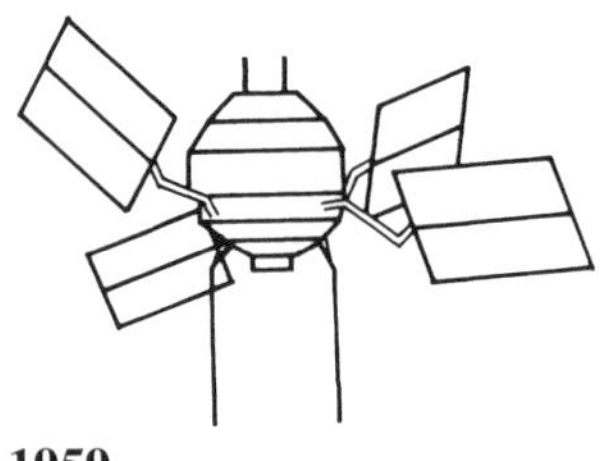

1959
Part of Explorer 6.

1963 • Discovery of the phenomenon of periodic magnetic reversals in the Earth's crust, supporting the theory of seafloor spreading and plate tectonics

1965 • Scientists speculate on the existence of "hotspots," junctures at tectonic plates through which heat leaks up into the ocean

1966 • Richard G. Doell, G. Brent Dalrymple, and Allan Cox demonstrate that the Earth's magnetic field has undergone periodic reversals between the North and South Poles

1968 • Evidence of life found in rocks dating back 3 billion years

1969 • First human being to set foot on the Moon is American astronaut Neil Armstrong

1970 • Celebration of the first Earth Day, marking the development of the "green movement," in which concerns are expressed over environmental damage

1972 • First American *Landsat* satellite is launched to study the Earth, including its mineral and agricultural resources

1974 • Damage to the Earth's ozone layer by chlorofluorocarbons (CFCs) discovered. Remains of human ancestor 3 million years old ("Lucy") discovered by American anthropologist Donald Johanson

1977 • Discovery of deep ocean vents of hot, mineral-laden water

1978 • U.S. satellite launched to study the Earth's oceans. 3.7-million-year-old footprints a probable human ancestor found in Tanzania by Anglo-Kenyan scientist Mary Leakey

1980 • U.S. satellite maps the Earth's magnetic field

1980s • Warmest decade since recording began in the 19th century. Richard Leakey and Alan Walker find an almost complete *Homo erectus* skeleton

1981 • First experimental work is carried out in Ocean Acoustic Tomography (CT scanning) to investigate below-the-surface features of oceans

1985 • A "hole" is detected in the ozone layer above Antarctica

1969
First man on the Moon.

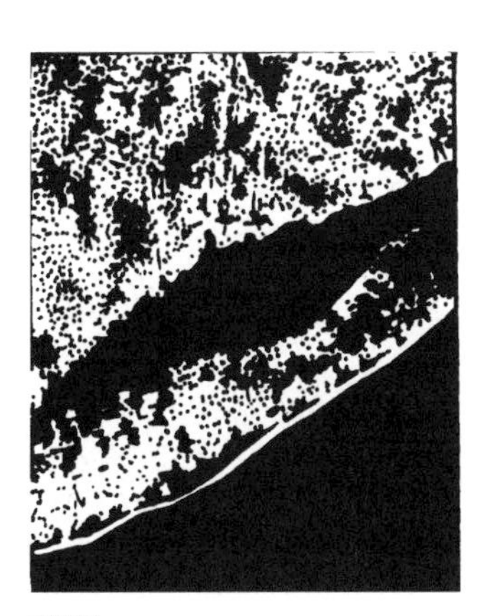

1972
View of New York from Landsat *satellite.*

1985
Hole in ozone layer discovered.

1988 • Fossil remains discovered by Israeli and French archaeologists prove that true *Homo sapiens* has been on Earth for more than 90,000 years, twice as long as previously thought

1989 • Many of the world's nations agree to plans to phase out ozone-destroying chlorofluorocarbons (CFCs). Discovery of the oldest known rock, dated at 3.96 billion years old. Astrophysicists discover the "Great Wall," a conglomeration of galaxies and the largest structure in the known universe

1990 • Discovery of the first freshwater geothermal vents

1991 • New seismic-wave travel timetables are developed that supersede the Jeffreys and Bullen tables. U.S. geologists confirm the theory that a large object from outer space smashed into the Earth 65 million years ago. Project *Biosphere 2* begins, in which eight people are locked in a sealed structure containing five sample Earth environments in order to study the feasibility of sustaining a closed ecology

1992 • Earth Summit conference held in Brazil to discuss climate change and various environmental issues

1993 • Remains of a 4-million-year-old homonid and probable human ancestor found in Ethiopia

1994 • Evidence of earliest known land life discovered dating from 1.2 billion years ago

1997 • At U.N. meetings in Kyoto, Japan, an agreement is signed by 5,000 representatives from 170 countries to take action on global warming by reducing carbon dioxide emissions into the atmosphere

1998 • Hurricane Mitch roars in across the southwest Caribbean and smashes into Honduras and Nicaragua. Winds in excess of 290 km an hour cause an estimated 11,000 deaths and billions of dollars of damage to villages, road and rail networks, water supplies, and farmland

1999 • Images from NASA's Chandra X-ray Observatory reveal the aftermath of a gigantic stellar explosion, including a possible neutron star or black hole

1998
The structure of a hurricane like Hurricane Mitch.

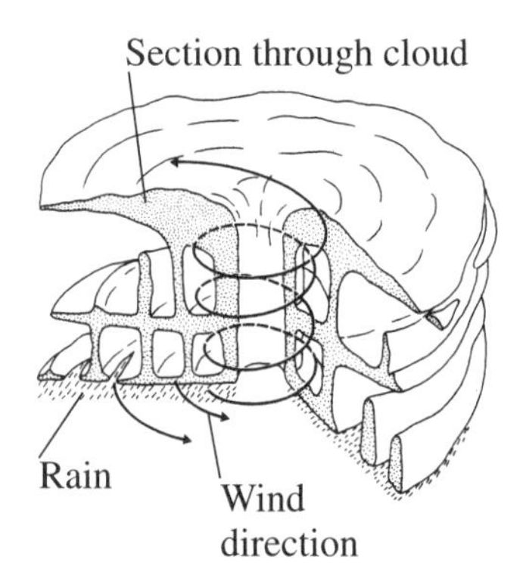

SECTION FOUR
CHARTS & TABLES

Geochronological time scale

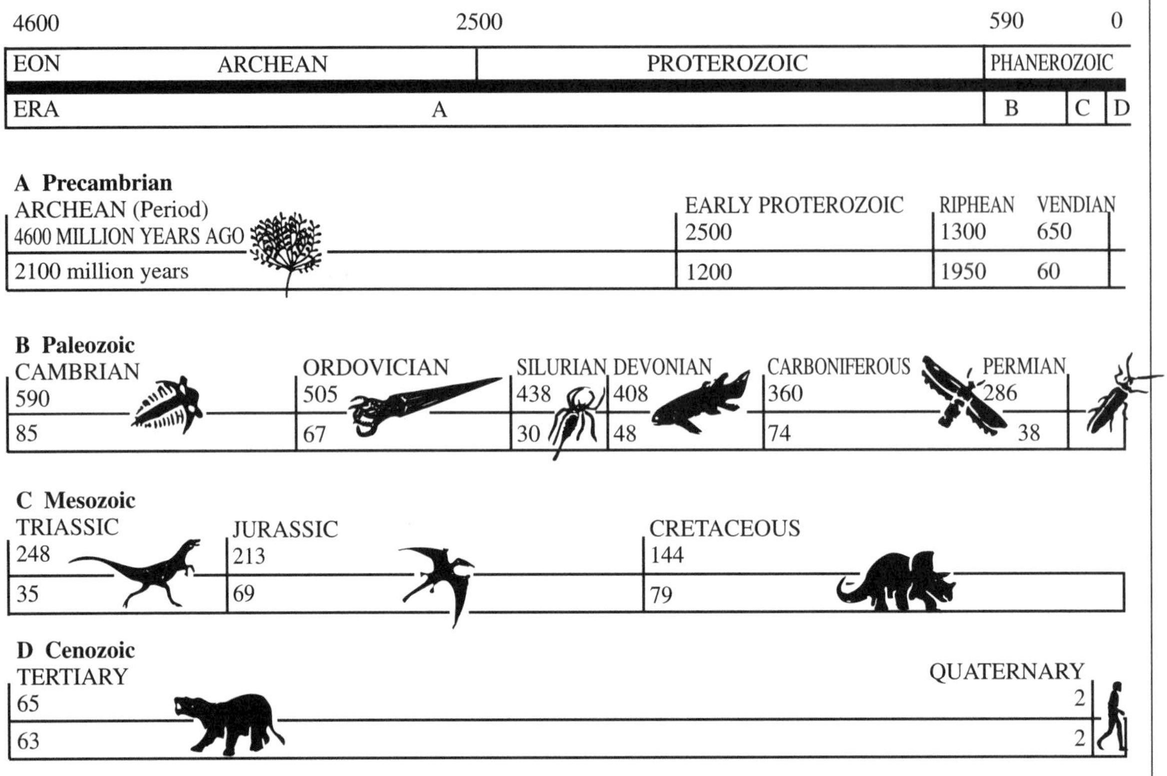

	4600	2500	590	0
EON	ARCHEAN	PROTEROZOIC	PHANEROZOIC	
ERA	A		B	C, D

A Precambrian

ARCHEAN (Period)	EARLY PROTEROZOIC	RIPHEAN	VENDIAN
4600 MILLION YEARS AGO	2500	1300	650
2100 million years	1200	1950	60

B Paleozoic

CAMBRIAN	ORDOVICIAN	SILURIAN	DEVONIAN	CARBONIFEROUS	PERMIAN
590	505	438	408	360	286
85	67	30	48	74	38

C Mesozoic

TRIASSIC	JURASSIC	CRETACEOUS
248	213	144
35	69	79

D Cenozoic

TERTIARY	QUATERNARY
65	2
63	2

Time periods

PRECAMBRIAN EONS (4600–591)

Archean

During this earliest and longest unit of geological time, the first small continents formed, volcanoes erupted on the early active surface, and the first bacteria appeared.

Proterozoic

As the Earth's crust cooled, larger continents evolved, mountains rose, and their eroded sediments accumulated below the sea. Complex living cells gave rise to simple multicellular organisms.

PALEOZOIC PERIODS (590–249)

Cambrian

Most continents, including a southern supercontinent, lay near the equator. Shallow seas teemed with early complex life forms, such as brachiopods, gastropods, graptolites, and trilobites.

Ordovician

A shrinking pre-Atlantic ocean brought proto-North America, Greenland, and Europe close. Ice covered some southern lands. Coral, dolomite, and limestone covered the shallow seafloor, and the first fish-like vertebrates appeared.

Silurian

Colliding northern proto-continents thrust up a mountain range from Scandinavia, through Scotland, to the Appalachians. Eroded debris formed thick sea sediments. First land plants (spore-bearing) established.

continued

Time periods continued

Devonian
Sandstone formed from the eroding arid Old Red Continent (eastern North America and Greenland fused with western Europe). Fish abounded, the first amphibians appeared with spiders and ammonites, and forests formed.

Carboniferous
Limestone formed below a shallow North American sea, followed by warm, swampy coal forests, inhabited by early reptiles and winged insects.

Permian
All the continents lay forced together as Pangaea. In arid inland areas, salt lakes produced evaporites and desert sandstones were formed. The drying up of shallow seas contributed to mass extinctions, but reptiles, beetles and coniferous trees survived.

MESOZOIC PERIODS (248–66)

Triassic
Pangaea showed signs of breaking up. Lands were mild or warm and largely dry. The dinosaurs, pterosaurs, and crocodilians all evolved from other reptiles in the group known as "archosaurs" ("ruling reptiles").

Jurassic
The Atlantic was opening up. The Tethys Sea divided northern and southern supercontinents, Laurasia and Gondwana, which were already splitting up into the continents we know today, inhabited by birds and mammals.

Cretaceous
Thick chalk deposits formed below shallow seas, covering parts of North America and Europe. Continental drift was under way and climates cooled. Dinosaurs and pterosaurs died out.

CENOZOIC PERIODS (65–2)

Tertiary
India merged with Asia, and colliding plates thrust up the Rockies, Alps, and Himalayas. Birds and mammals evolved and multiplied to occupy the gaps left by the vanished dinosaurs and pterosaurs. Flowering plants now dominated other kinds. There was great diversity of plant and animal life, culminating in the earliest hominids.

Quaternary
As temperatures dropped, ice sheets covered Antarctica and large parts of the Northern Hemisphere, and the ocean level fell. In warm phases, ice retreated and the ocean level rose. The present warm interval began about 10,000 years ago. Development of primates and early hominids toward modern human beings.

Structure of the Earth

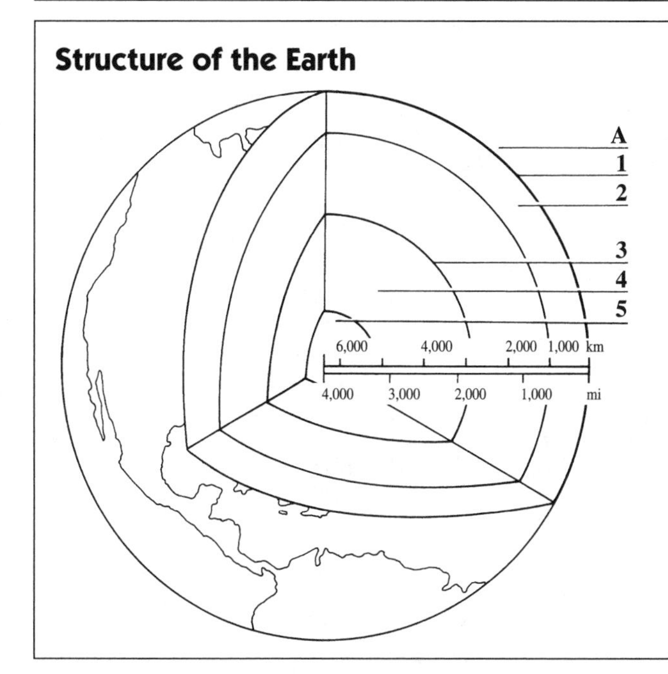

A Atmosphere
1 Crust
2 Upper mantle
3 Gutenburg (Oldham) Discontinuity
4 Outer zone
5 Inner zone

Origin and structure of the Earth

The Earth may have formed in the following four stages:
1 Cloud of chemical particles.
2 Densest particles sink inward.
3 Continued sorting of particles leads to primeval Earth:
a Dense core of iron and nickel
b Less dense matter similar to that of the meteorites (carbonaceous chondrites)

4 Formation of the Earth's major layers:
a Dense core
b Less dense mantle (melted chondite)
c Crust
d Ocean
e Primeval atmosphere

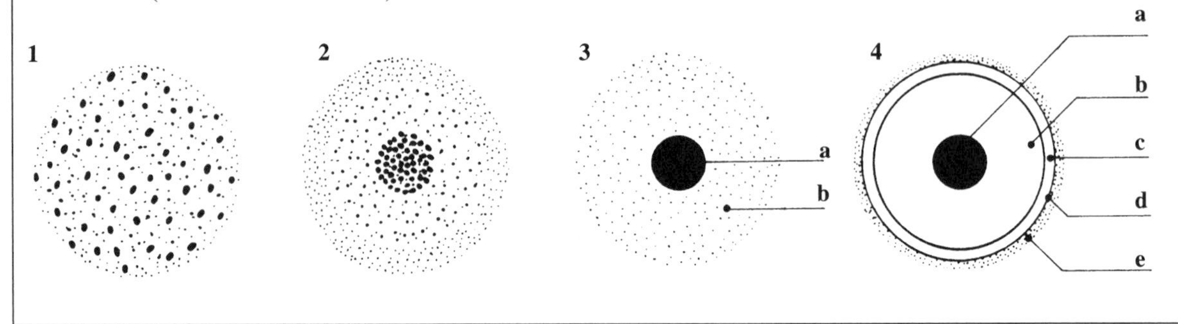

Evolving continents

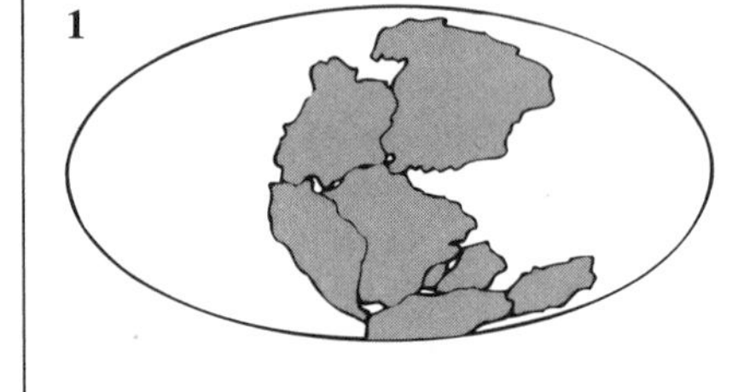

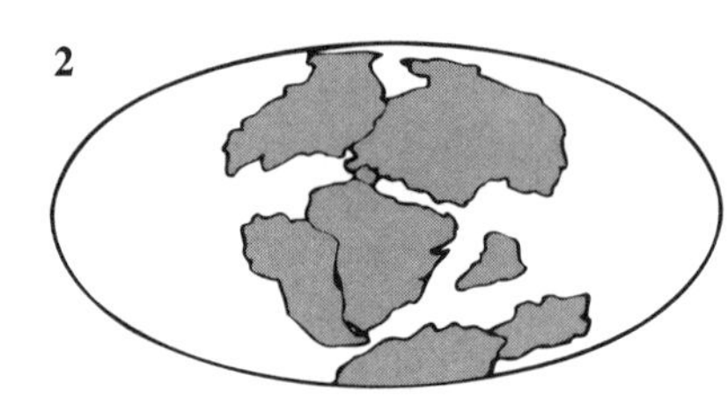

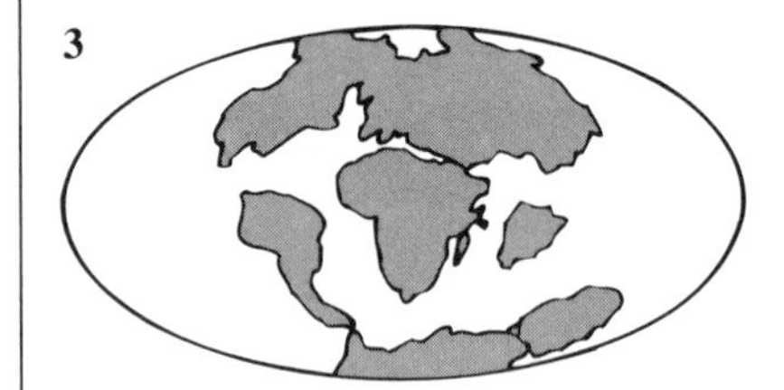

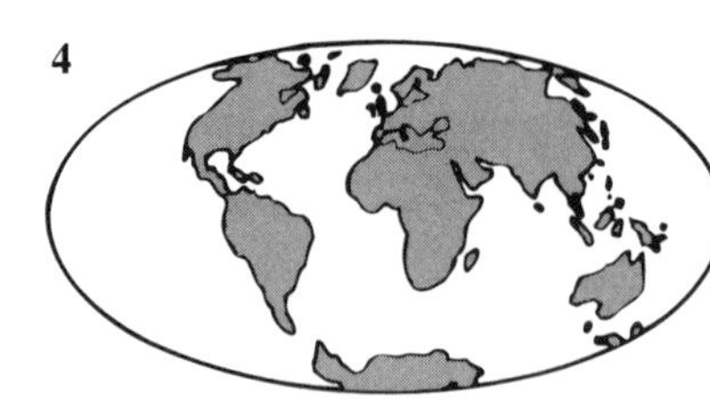

1 200 million years ago, the supercontinent Pangaea ("All Earth") was breaking up and its components began to drift.
2 140 million years ago, break-up had produced a northern landmass, Laurasia, and a southern landmass, Gondwana, separated by the Tethys Sea.
3 65 million years ago, the widening Atlantic Ocean had separated the Americas from Africa, and continents were gaining their present shapes and positions. India had broken free from Africa but not yet docked with Asia.
4 The current shapes and positions of the seven continents. The Atlantic Ocean is still widening.

Rock cycle

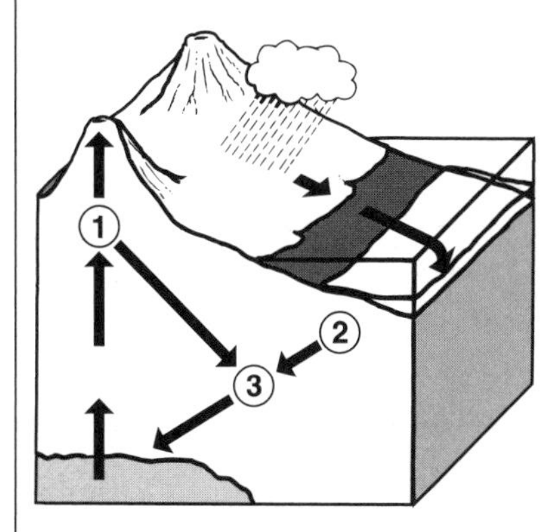

The rock cycle is a continuous round of erosion, deposition, and heat.
1 Igneous (volcanic) rocks are formed when magma (liquid rock) is forced to the surface under pressure and erupts through a volcano. As the magma reaches the surface, it solidifies and becomes igneous rock.
2 The rock is now exposed to the elements, and over a long period of time begins to be eroded and weathered. The rock particles are carried by rivers, wind, and rain to the oceans and seas. The particles build up into layers that over many thousands of years are buried deeper and deeper. The weight of layers presses the lowest layers into a solid; this is called sedimentary rock.
3 The constant pressure pushes the sedimentary rock further and further down, where it begins to be affected by heat from the Earth's core. This heat causes the structure of the rock to change, or metamorphose, and it now becomes known as metamorphic rock. If it remains under high pressure and constant heat, the rock will eventually become magma, which may rise to the surface under pressure and erupt again as a volcano.

Igneous rocks are formed when magma from the inside of the Earth rises to the surface and cools. The size of the crystals in the rock shows the speed at which the magma cooled. Granite and basalt are igneous rocks.
1 Pumice is made by gases bubbling through lava.
2 Basalt hardens quickly above the surface. The rock crystals are small because they have not had much time to grow.
3 Granite hardens slowly below the surface; the crystals are large.

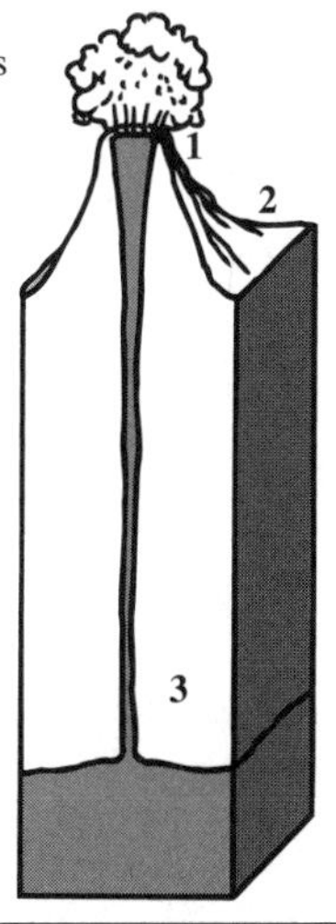

Sedimentary rocks are mainly made from the remains, or sediments, of older rocks that have been worn away. These are washed into rivers and out to sea. Fragments settle near the shore, forming sandstone (**1**). Further offshore, rock deposits collect to form shale (**2**), and in warm seas, deposits collect to form limestone (**3**). On the seabed the sediments are squeezed together, eventually becoming solid rock.

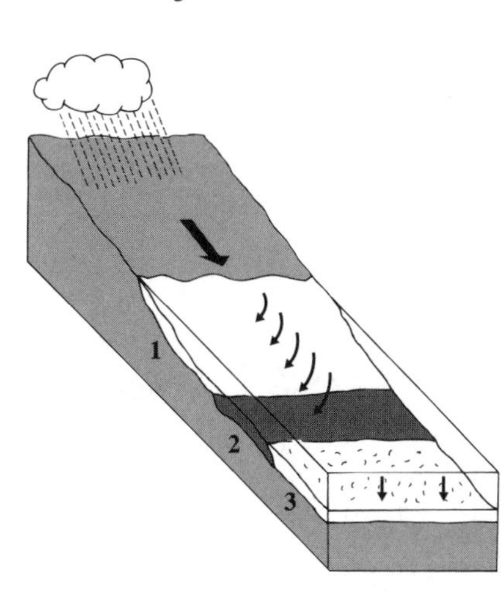

Metamorphic rocks are formed from other rocks that have been heated or placed under pressure. These factors change the rock. Metamorphic rocks such as marble or slate tend to have crystals in bands, or layers.

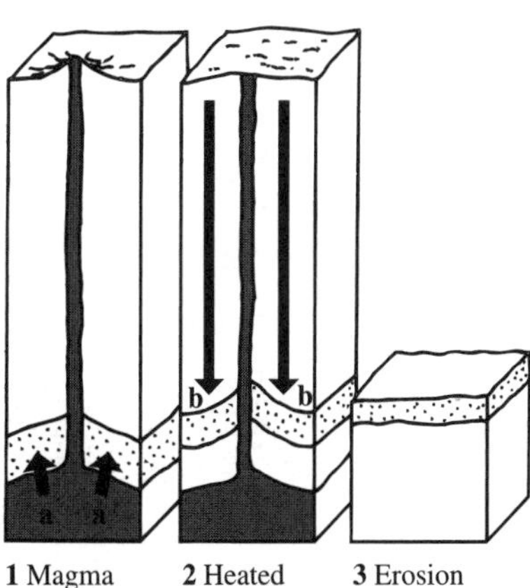

1 Magma heats (**a**) limestone
2 Heated limestone is compressed (**b**)
3 Erosion reveals marble

Earth plate tectonics

The Earth's surface consists of large pieces, or plates, that fit together rather like a jigsaw. They form the lithosphere, or the Earth's outer shell. The plates are constantly moving. Plate tectonics is the study of how they move. When plates slide past each other, they cause fractures or faults. If the rock bends instead of breaking, it creates folds.

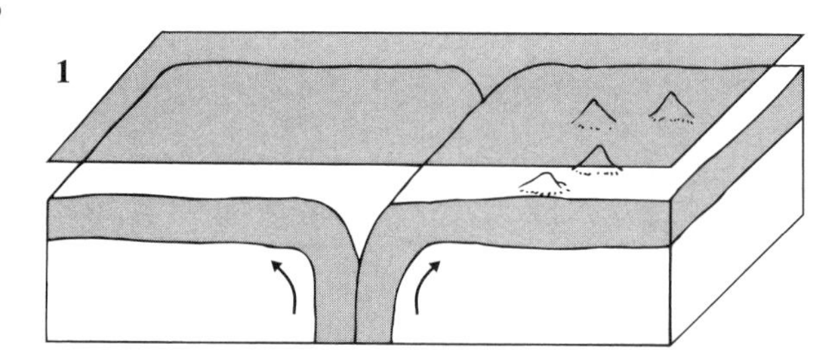

1 Spreading
Two plates spreading apart form ocean floors and underwater volcanoes

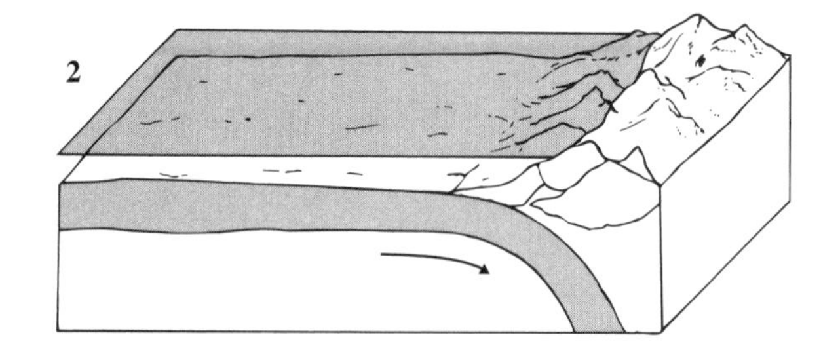

2 Subduction
Plates that push against each other and bend downward are called subduction zones

Lithospheric plates
- **1** African plate
- **2** Antarctic plate
- **3** Arabian plate
- **4** Caribbean plate
- **5** Eurasian plate
- **6** Indo-Australian plate
- **7** Nazca plate
- **8** North American plate
- **9** Pacific plate
- **10** Philippines plate
- **11** South American plate

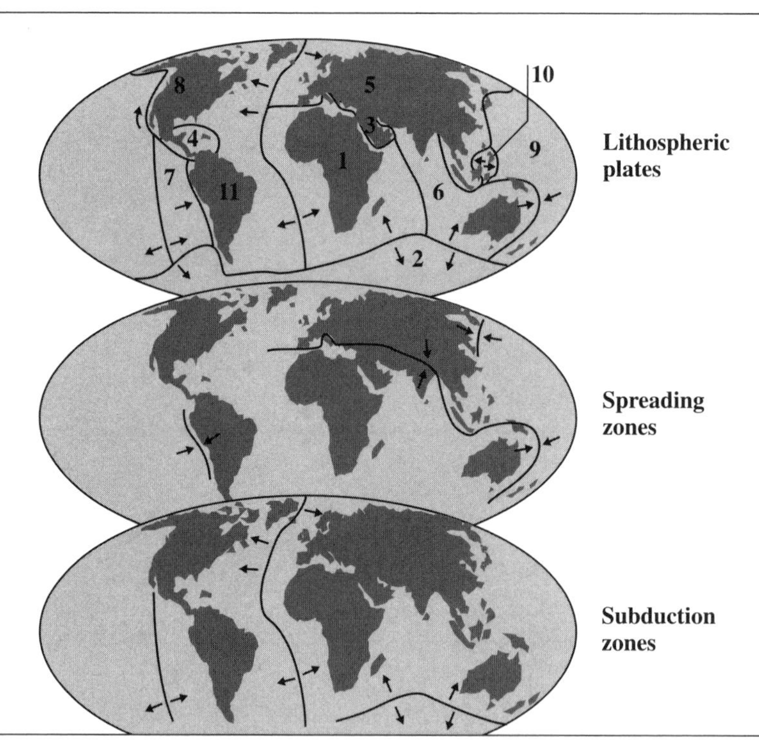

Lithospheric plates

Spreading zones

Subduction zones

Folds and faults

Folds
Folds are the buckling of once horizontal rock strata, frequently caused by rocks being crumpled at plate boundaries.

a) Anticline fold Pressure from beneath forces the Earth's surface upward.
b) Syncline fold Activities within sedimentary layers allow the surface to sink.
c) Over fold Sideways pressures force the surface to buckle and fold over itself.
d) Nappe fold Extreme overfolding shears rocks that ride over subterranean layers.

a

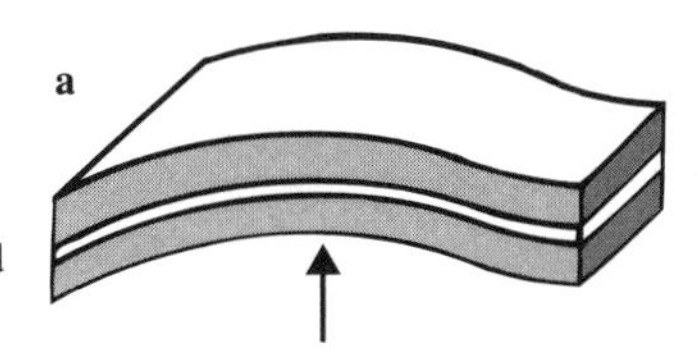

b

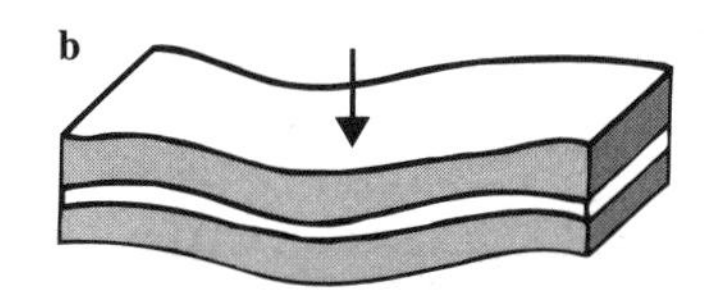

c

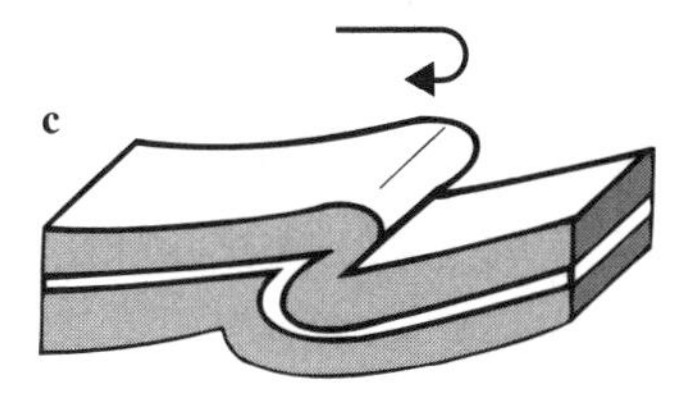

d

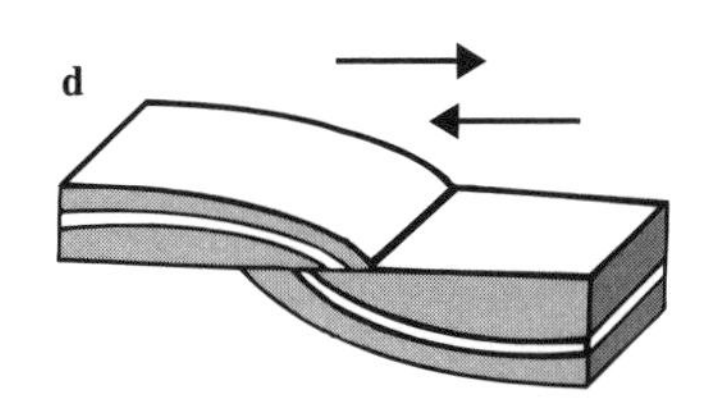

Faults
Faults are fractures in the Earth's crust either side of which rocks have been moved relative to each other.

e) Normal fault Stretching surfaces cause rocks to move upward and downward to reveal an escarpment (exposed new surface).
f) Reverse fault Compression of an area causes rocks to ride up over others to produce an overhang.
g) Tear fault Sideways-moving plates that produce earthquakes.
h) Graben A long narrow area that sinks between two parallel faults.
j) Horst A horizontal block raised between two normal faults.

e

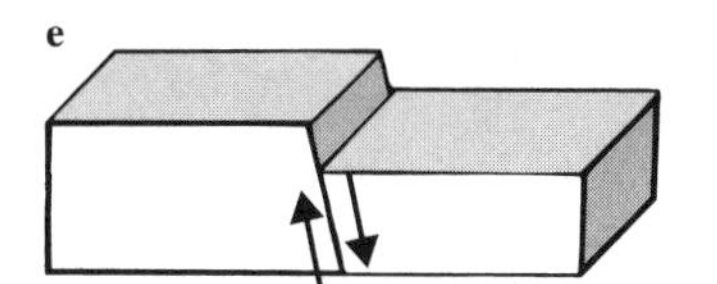

f

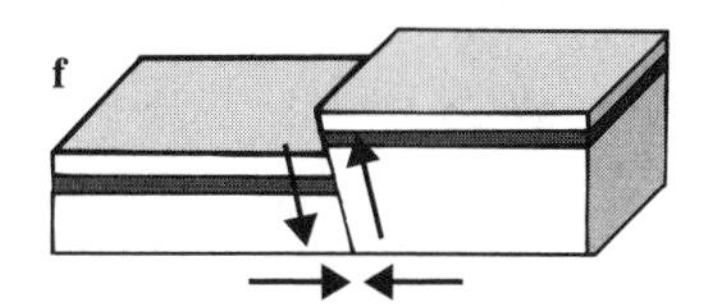

g

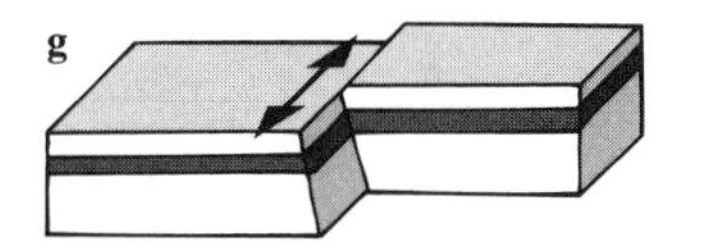

h

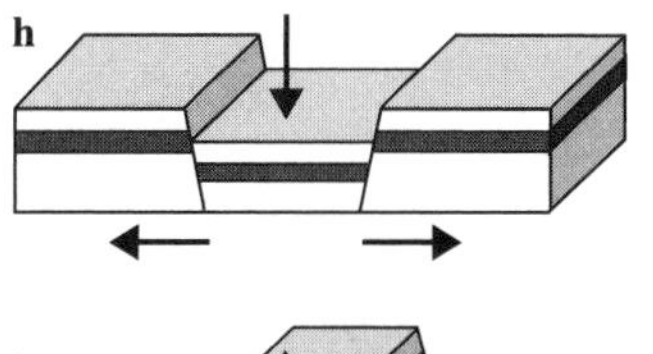

j

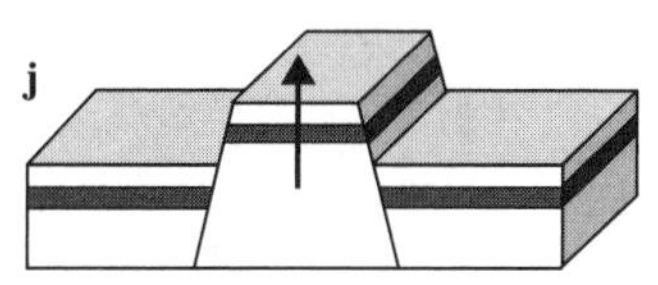

Mercalli and Richter scales for earthquakes

Mercalli number	Intensity	Effects	Richter number
I	Instrumental	Detected by seismographs and some animals.	< 3.5
II	Feeble	Noticed by a few sensitive people at rest.	3.5
III	Slight	Similar to vibrations from a passing truck.	4.2
IV	Moderate	Felt generally indoors; parked cars rock.	4.5
V	Rather strong	Felt generally; most sleepers wake.	4.8
VI	Strong	Trees shake; chairs fall over; some damage.	5.4
VII	Very strong	General alarm; walls crack; plaster falls.	6.1
VIII	Destructive	Chimneys, columns, monuments, weak walls fall.	6.5
IX	Ruinous	Some houses collapse as ground cracks.	6.9
X	Disastrous	Many buildings destroyed; railroad lines bend.	7.3
XI	Very disastrous	Few buildings survive; bad landslides and floods.	8.1
XII	Catastrophic	Total destruction; ground forms waves.	> 8.1

Folds and faults – Mercalli and Richter scales

CHARTS & TABLES

World distribution of mountains

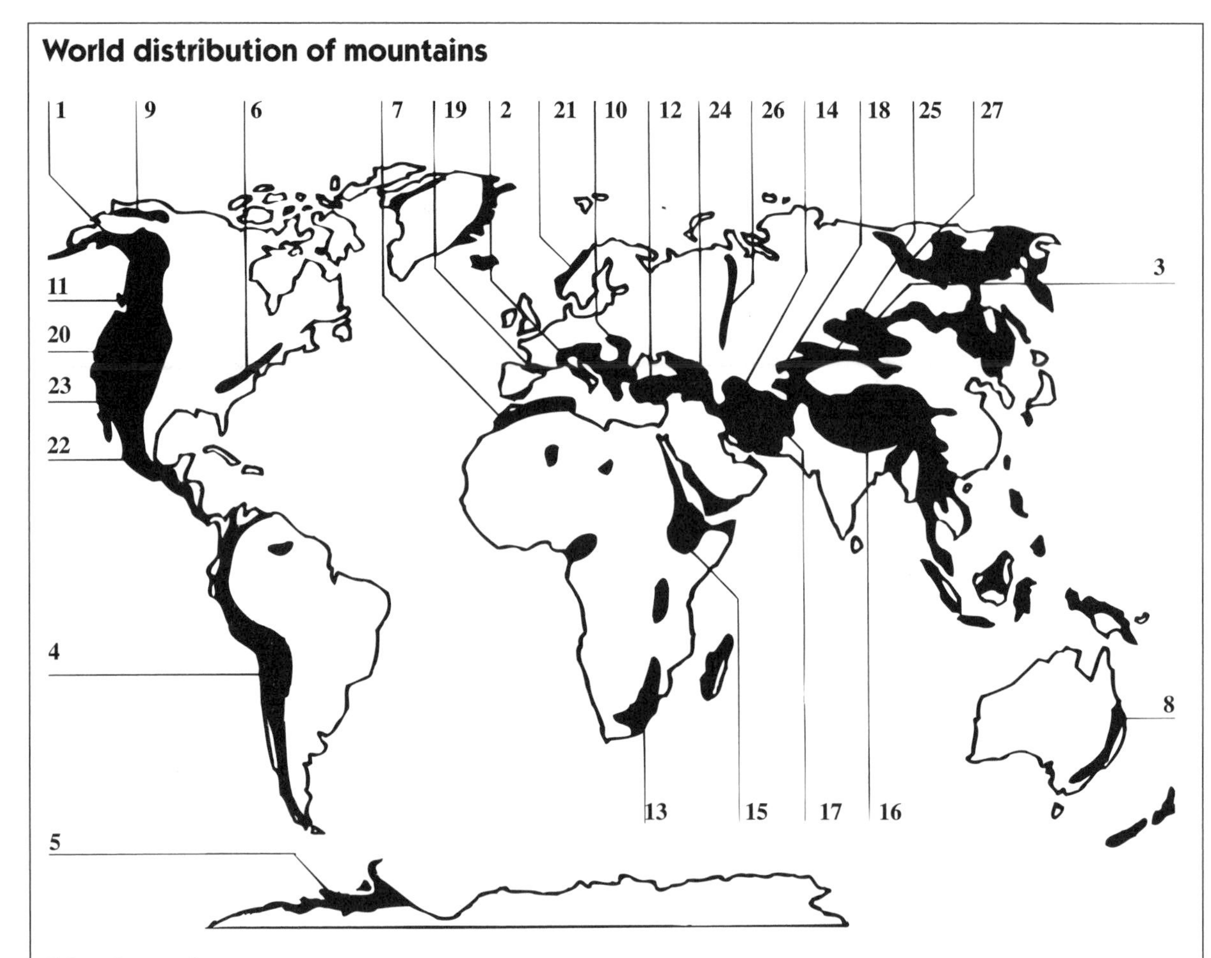

Selected mountain groups

The list gives the major mountain ranges of the world (shown above) and the highest peak in each of the ranges, with its height shown in thousands of feet (m).

1 Alaska Range: Mt McKinley 20.3 (6.2)
2 Alps: Mont Blanc 15.8 (4.8)
3 Altai Range: Gora Belucha 14.8 (4.5)
4 Andes: Aconcagua 22.8 (6.9)
5 Antarctic Peninsula: Vinson Massif 16.9 (5.1)
6 Appalachian Mountains: Mt Mitchell 6.6 (2.0)
7 Atlas Mountains: Jbel Toubkal 13.5 (4.2)
8 Australian Alps: Mt Kosciusko 7.3 (2.2)
9 Brooks Range: Mt Isto 8.9 (2.8)
10 Carpathian Mountains: Moldoveanul 8.2 (2.5)
11 Cascade Range: Mt Rainier 14.2 (4.4)
12 Caucasus Mountains: Mt Elbrus 18.5 (5.6)
13 Drakensberg Mountains: Thabana Ntlenyana 11.4 (3.5)
14 Elburz Mountains: Mt Demavend 18.4 (5.6)
15 Ethiopian Highlands: Ras Dachan 15.2 (4.6)
16 Himalaya: Mt Everest 29 (8.8)
17 Hindu Kush: Tirich Mir 25.2 (7.7)
18 Pamirs: Pik Kommunizma 24.6 (7.5)
19 Pyrenees: Pic d'Aneto 11.2 (3.4)
20 Rocky Mountains: Mt Elbert 14.4 (4.4)
21 Scandinavian Mountains: Glittertind 7.9 (2.4)
22 Sierra Madre: Citlaltépetl 18.7 (5.7)
23 Sierra Nevada: Mt Whitney 14.5 (4.4)
24 Taurus Mountains: Aladag 12.2 (3.7)
25 Tien Shan: Pik Pohedy 24.4 (7.4)
26 Ural Mountains: Mt Narodnaya 6.2 (1.9)
27 Verkhoyansk Range: Gora Pobeda 10.3 (3.1)

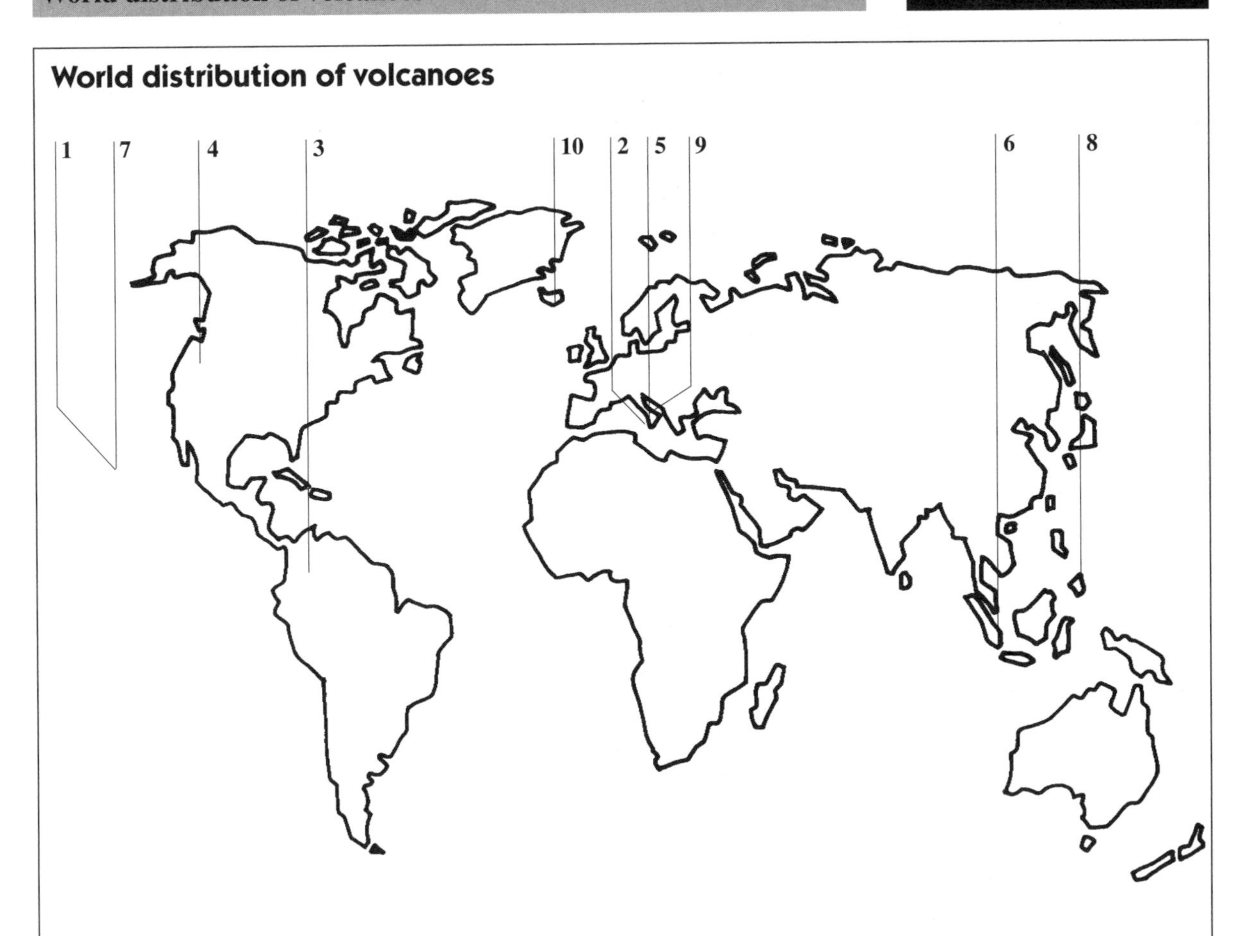

Sometimes holes and cracks appear in the Earth's crust. These release hot liquid rock, called magma, ash, gases, and other matter. As the magma and ash cool, they harden to form rock. Generally, volcanoes erupt when the Earth's plates move apart or when one plate is forced beneath another. Throughout the world there are 500 to 800 active volcanoes. However, during a year only 30 might erupt on land. An inactive, or dormant, volcano is still capable of erupting. An extinct volcano is dead and cannot erupt.

Selected active volcanoes

Volcano	Country	Major periods of eruption
1 Kilauea	USA (Hawaii)	1823–present
2 Stromboli	Italy	1768–1989
3 Nevado del Ruiz	Colombia	1985
4 St Helens, Mt	USA	1800–87, 1989
5 Mt Etna	Italy	1947–1991
6 Krakatoa	Sumatra	1680–1972
7 Mauna Loa	USA (Hawaii)	1859–1987
8 Mt Pinatubo	Philippines	1380–1991
9 Vesuvius	Italy	CE 79–1944
10 Hekla	Iceland	1693–1970

World climates

1 Polar climate zone
2 Temperate climate zone
3 Tropical climate zone

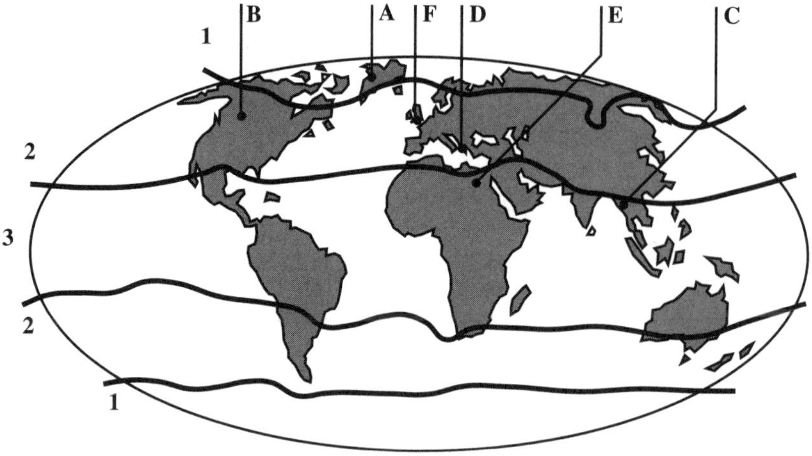

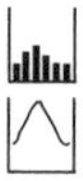
Rainfall (mm)
Temperature °C

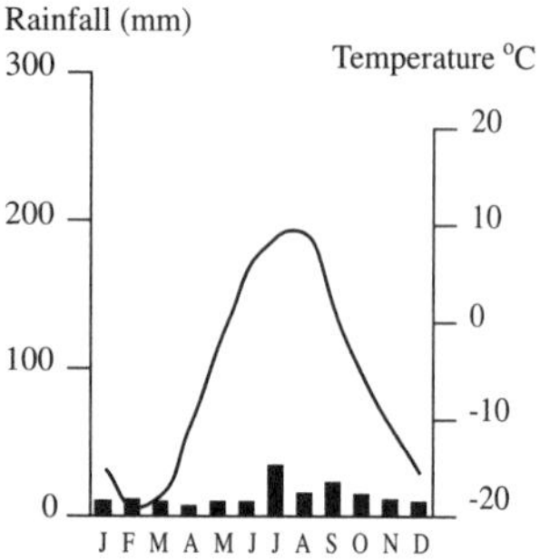

A Polar
Thule (Greenland) Total 93 mm

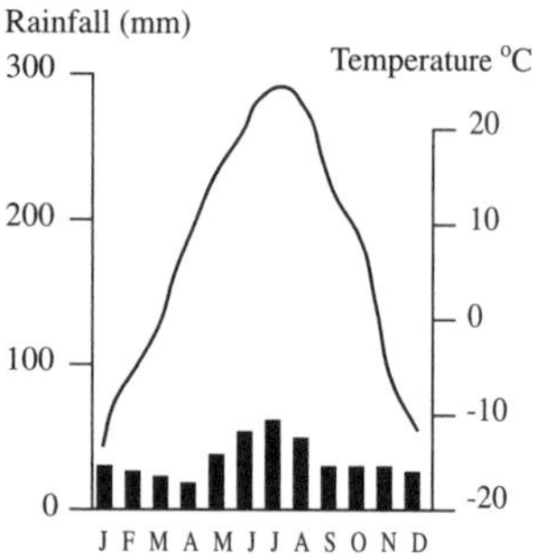

B Cold temperate (continental)
Peace River (Canada)
Total 376 mm

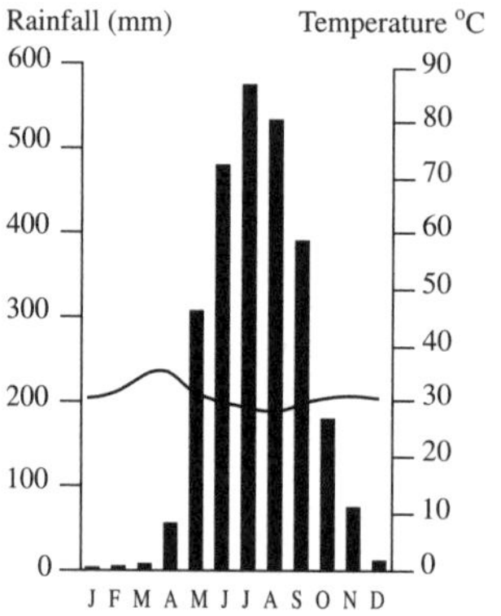

C Tropical (monsoon)
Yangon (Myanmar) Total 2,620 mm

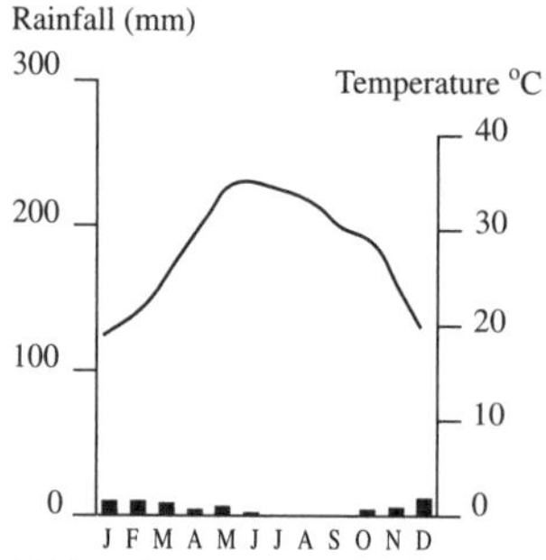

E Tropical (desert)
Cairo (Egypt) Total 25 mm

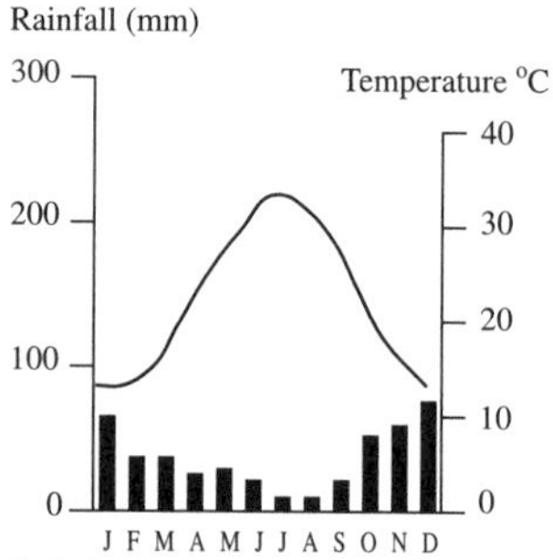

D Warm temperate
Athens (Greece) Total 402 mm

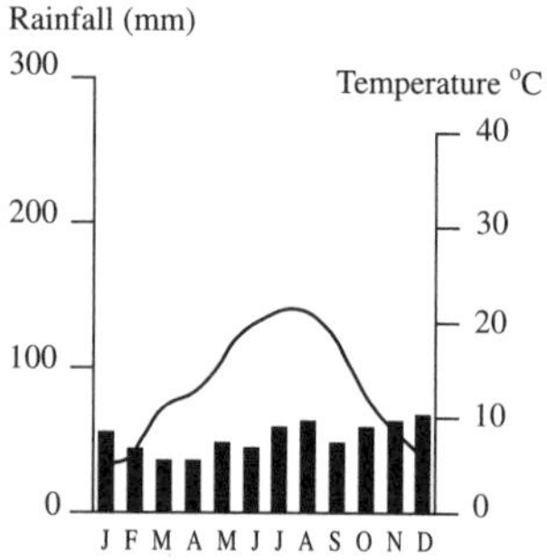

F Cool temperate (marine)
London (UK) Total 593 mm

CHARTS & TABLES World climates

World wind systems

1

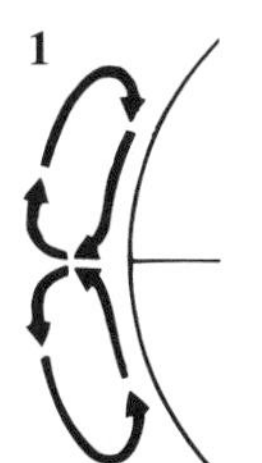

2

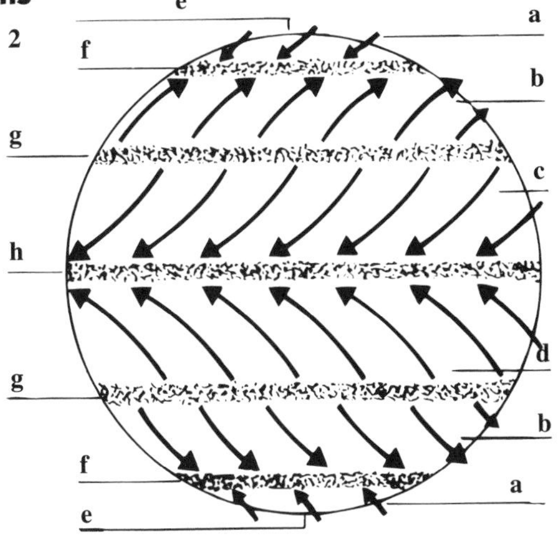

Surface winds
a Polar easterly winds
b Midlatitude westerly winds
c Northeast trade winds
d Southeast trade winds

Surface pressure belts
e Polar high pressure
f Temperate low pressure
g Horse latitudes
h Doldrums

January wind pattern

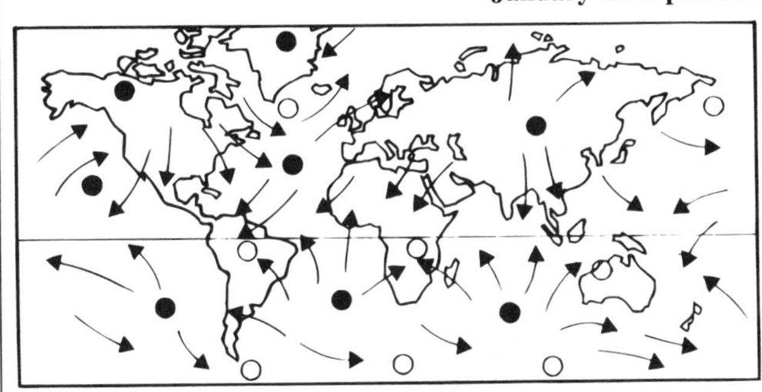

July wind pattern

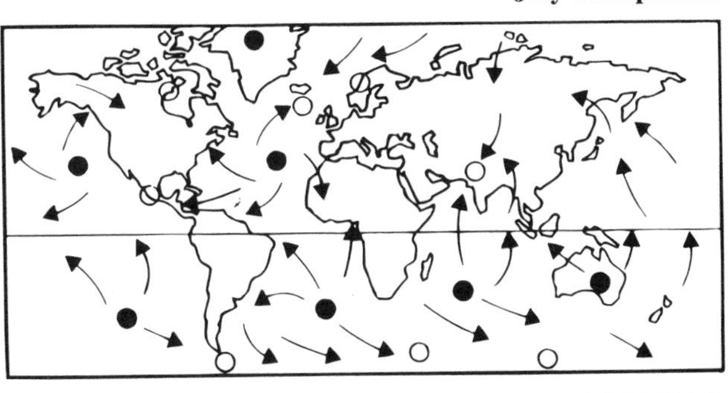

The Sun's heat and the Earth's spin produce the large-scale atmospheric circulation that we call the world's prevailing winds.

1 Hot air rises from the doldrums near the Equator and thins out. It spreads north and south, cools down, and sinks in the horse latitudes, and becomes dense. It then flows back toward the Equator.
2 Prevailing surface winds (a–d) reflect the atmospheric flow in 1 that produces the surface pressure belts (e–g). Dense air from the polar highs and the high-pressure horse latitudes flows toward the temperate low-pressure belts and the low-pressure doldrums.

The maps show wind patterns for January and July and illustrate some interesting features of global wind movement. Winds blowing over North America and Asia are out-blowing during the winter when pressure is high but are inblowing during the summer when pressure is low.

Monsoon winds
The word *monsoon* is derived from the Arabic word *mausin*, and means "season." It is always used when describing winds whose direction is reversed completely from one season to the next; these most often develop over Asia.

● high pressure
○ low pressure

World ocean currents

Ocean currents are the movements of the surface water. The most important causes are the prevailing winds and differences in water density due to temperature or salinity. The shape of the continents and the rotation of the Earth can also influence the direction of currents. Ocean currents caused by prevailing winds are called drift currents, the best known being the Gulf Stream. Between the Equator and the temperate regions in the Northern Hemisphere (**1**), the circulation of ocean currents is clockwise, in the Southern Hemisphere (**2**) it is counterclockwise (see below). In equatorial regions, currents move in opposite directions, those in the north moving left to right, those in the south from right to left (**3**). Currents moving north and south from equatorial regions carry warm water and those moving south and north from polar regions carry cold water.

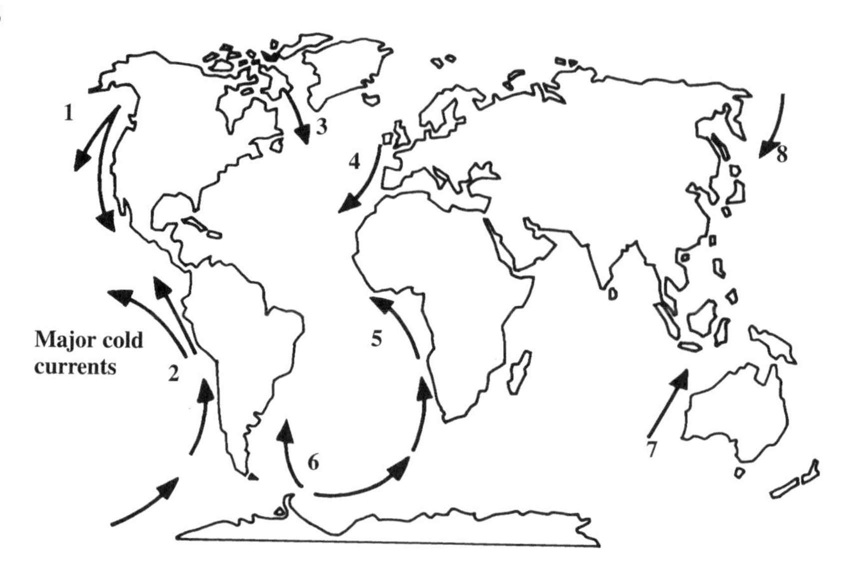

Major cold currents (above)
1 California
2 Humboldt
3 Labrador
4 Canaries
5 Benguela
6 Falkland
7 West Australian
8 Okhotsk

Major warm currents (below)
1 North Atlantic (Gulf Stream)
2 South Atlantic
3 South Indian Ocean
4 South Pacific
5 North Pacific
6 Monsoons

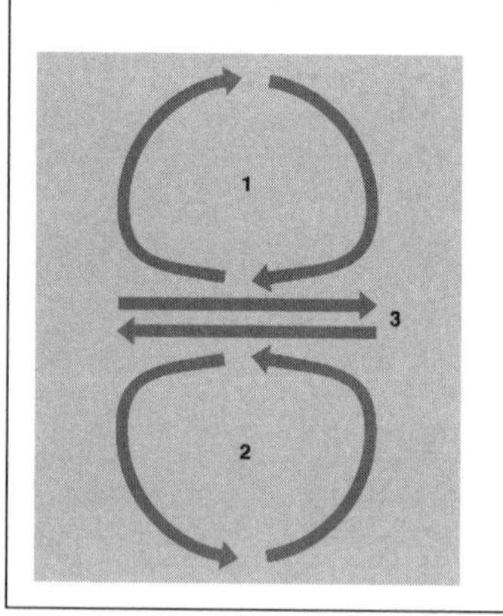

Weather map symbols

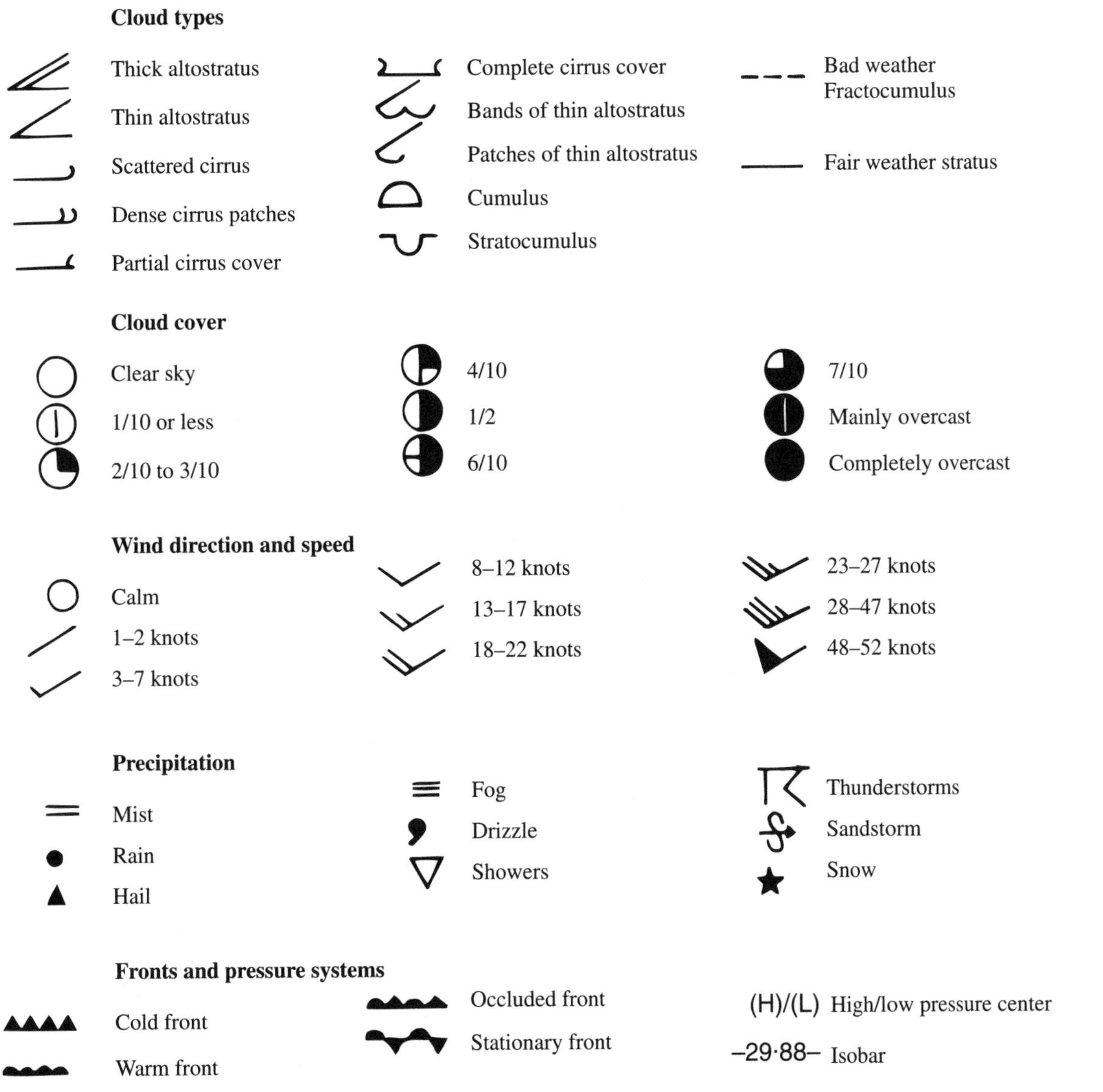

Cloud types

Thick altostratus
Thin altostratus
Scattered cirrus
Dense cirrus patches
Partial cirrus cover
Complete cirrus cover
Bands of thin altostratus
Patches of thin altostratus
Cumulus
Stratocumulus
Bad weather Fractocumulus
Fair weather stratus

Cloud cover

Clear sky
1/10 or less
2/10 to 3/10
4/10
1/2
6/10
7/10
Mainly overcast
Completely overcast

Wind direction and speed

Calm
1–2 knots
3–7 knots
8–12 knots
13–17 knots
18–22 knots
23–27 knots
28–47 knots
48–52 knots

Precipitation

Mist
Rain
Hail
Fog
Drizzle
Showers
Thunderstorms
Sandstorm
Snow

Fronts and pressure systems

Cold front
Warm front
Occluded front
Stationary front
(H)/(L) High/low pressure center
–29·88– Isobar

Cloud types

A Stratus: a low, gray layer; drizzle or snow grains.
B Cumulus: low, detached with dark level bases and white fluffy tops; may bring showers.
C Stratocumulus: low, whitish or gray with dark parts, wavelike or patchy with no rain.
D Cumulonimbus: towering, with a dark base and anvil-shaped top; thunderstorms.
E Nimbostratus: middle-altitude, dark, dense, often ragged beneath; rain or snow.
F Altostratus: middle-altitude, grayish or bluish sheets, rainbearing.
G Altocumulus: middle-altitude white or gray rolls, "mackerel sky."
H Cirrus: high-altitude, thin white, wispy; made of ice crystals.
I Cirrostratus: high, whitish, transparent.
J Cirrocumulus: high, white patches in sheets or layers; made up of ice crystals.

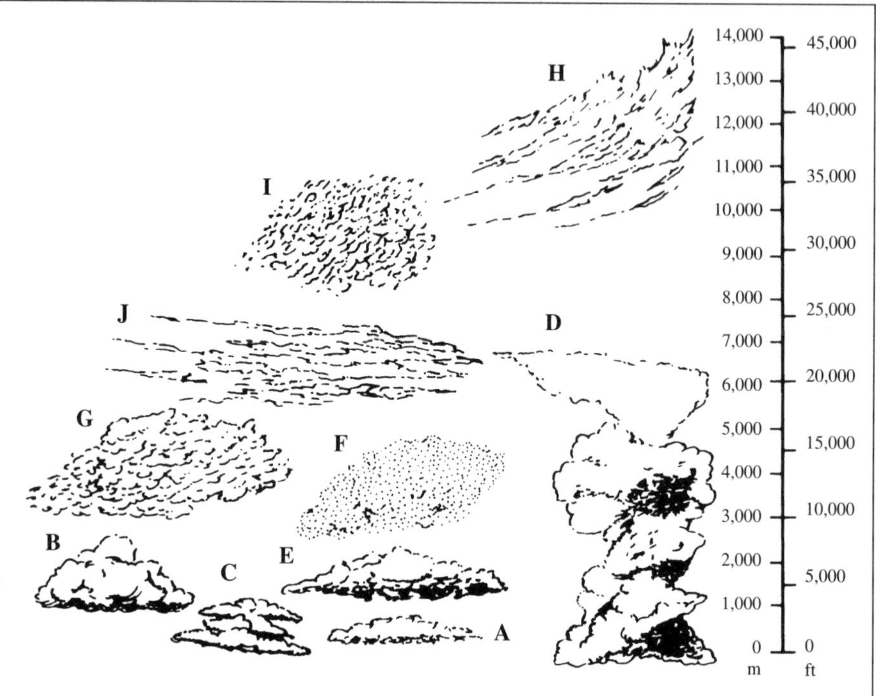

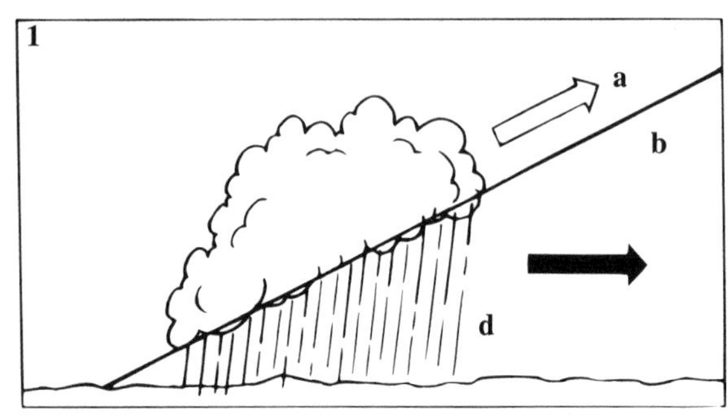

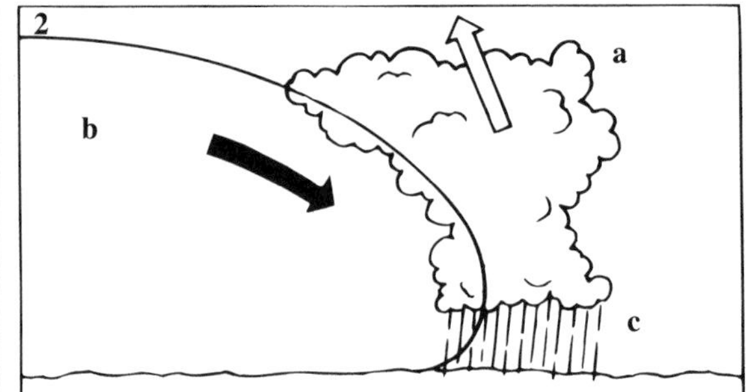

Lows (depressions) bring wind and rain to the temperate midlatitudes. Whirling "pinwheels" of air hundreds of miles across, they form where cold polar air clashes with warm, moist, subtropical air along a boundary, called the polar front. Lows bring warm and cold fronts.

1 At a warm front, warm, moist air rides up over colder air, producing sheetlike stratus cloud shedding steady drizzle or snow.

2 At a cold front following a warm front, cold air undercuts warmer air from behind. This may produce dark nimbostratus clouds shedding heavy showers of rain or snow.

a Warm air
b Cold air
c Heavy showers
d Prolonged drizzle

Winds

Number	Description	Speed (mi/hr)	(km/hr)	Characteristics
0	Calm	Below 1	(Below 2)	Smoke goes straight up
1	Light air	1–3	(2–5)	Smoke blown by wind
2	Light breeze	4–7	(6–11)	Wind felt on face
3	Gentle breeze	8–12	(12–19)	Extends a light flag
4	Moderate breeze	13–18	(20–29)	Raises dust and loose paper
5	Fresh breeze	19–24	(30–38)	Small trees begin to sway
6	Strong breeze	25–31	(39–50)	Umbrellas become hard to use
7	Moderate gale	32–38	(51–61)	Difficult to walk into
8	Fresh gale	39–46	(62–74)	Twigs broken off trees
9	Strong gale	47–54	(75–86)	Roof damage
10	Whole gale	55–63	(87–102)	Trees uprooted
11	Storm	64–73	(103–117)	Widespread damage
12–17	Hurricane	74 and up	(118 and up)	Violent destruction

The Beaufort scale
This is an internationally recognized scale for describing wind speeds that are 33 feet (10 m) above ground level. The table gives standard descriptions and wind speeds corresponding to each number in the Beaufort scale. It originated in 1805 when British admiral Sir Francis Beaufort devised a scale of numbers to describe the effects of winds of different speeds on sailing ships.

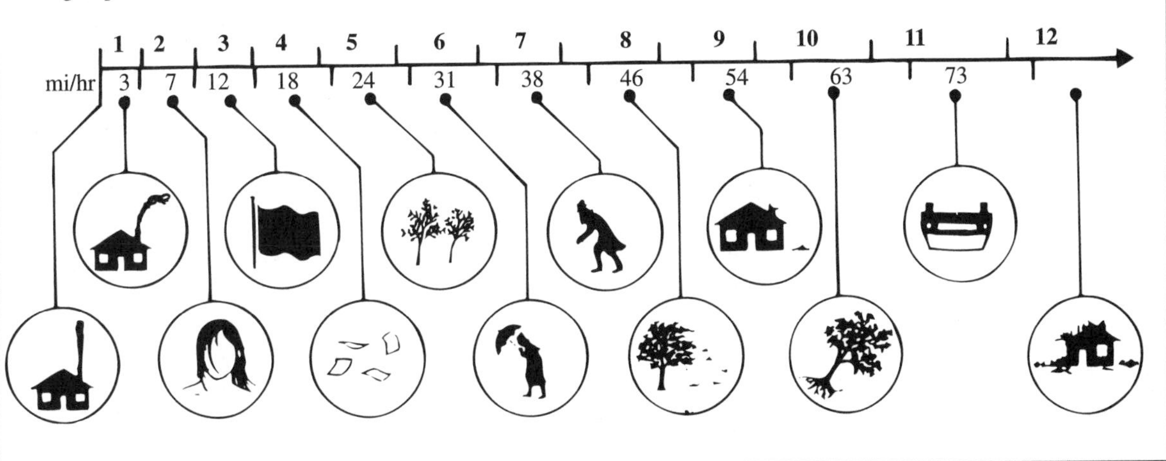

Topographic map symbols

- Primary highway (r)
- Secondary highway (r)
- Light-duty road
- Unimproved track
- Single-track railroad
- Multiple-track railroad
- Buildings
- Landmark; windmill
- Quarry; prospect
- Spot elevation
- Index contour
- Supplementary contour
- Intermediate contour
- Embankment

- Cutting
- National boundary
- State boundary
- Country parish boundary
- Perennial stream (b)
- Intermittent stream (b)
- Water well; spring (b)
- Small rapids (b)
- Big rapids (b)
- Big falls (b)
- Small falls (b)
- Intermittent lake (b)
- Glacier (w)
- Dry lake bed (r-in-b)

- Marsh or swamp (b-on-w)
- Wooded marsh (g)

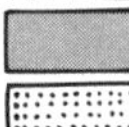

- Woods (g)
- Vineyard (g)
- Controlled flooding
- Submerged marsh (b)

- Mangrove swamp (g)

- Orchard (g-on-w)

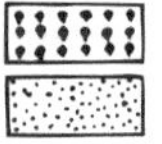

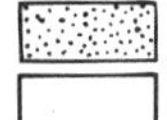

- Scrub (g-on-w)
- Urban area (p)

Colour key

red (r)
green (g)
blue (b)
pink (p)
white (w)
blue-on-white (b-on-w)
red-in-blue (r-in-b)
green-on-white (g-on-w)

Latitude and longitude

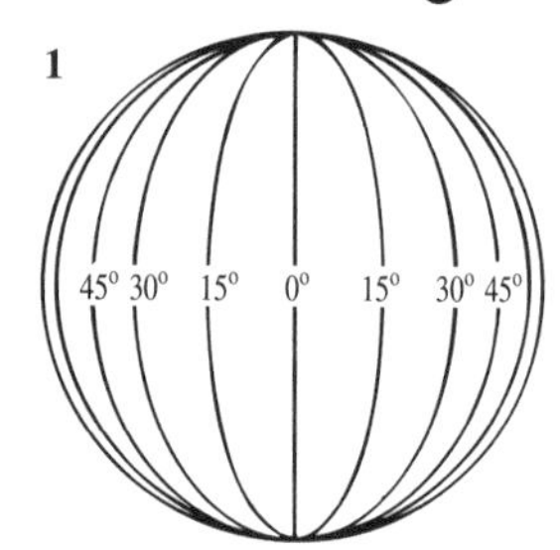

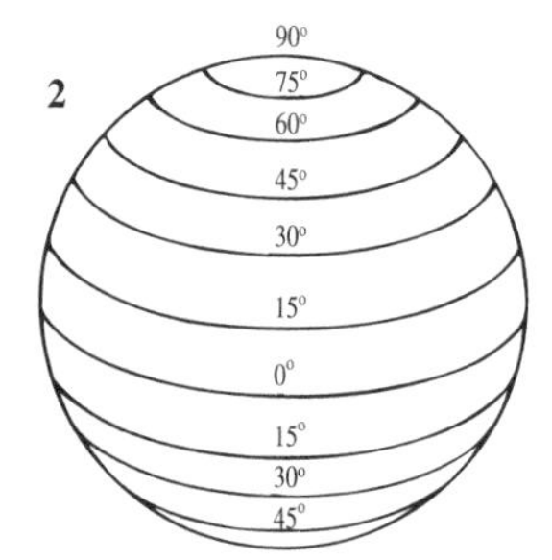

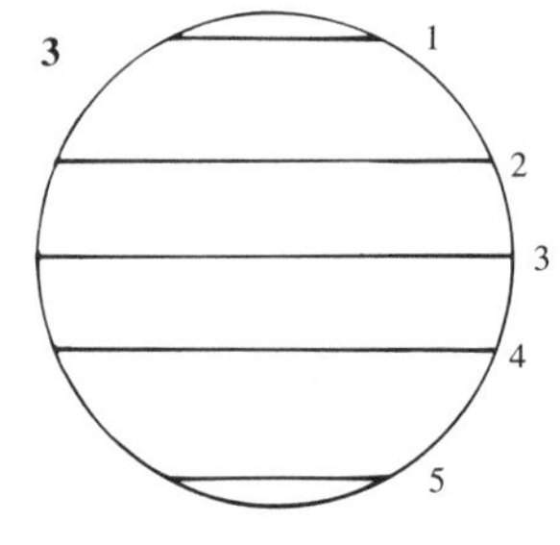

Longitude
On a map or globe, longitude is a position east or west of an imaginary north-south line between the North and South Poles. The prime meridian passes through Greenwich, England. Longitude measurements are given in degrees, minutes, and seconds, e.g. 3°08'24" W (west). (The distance between two degrees of latitude or longitude is 60 minutes, and that between two minutes of latitude or longitude is 60 seconds.) **1** labels the prime meridian (0°) and meridians at 15° intervals east and west of it.

Latitude
On a map or globe, latitude is a position north or south of the equator, an imaginary east-west line around the world halfway between the poles. Latitude measurements are given in degrees, minutes, and seconds – e.g. 52°51'02" N (north). **2** labels the equator (0°) and parallels at 15° intervals north and south. The North Pole is 90° N and the South Pole is 90° S.

Key latitudes (3)
Numbers indicate important latitudes. Between the tropics, the Sun shines down vertically at least once a year. North of the Arctic Circle and south of the Antarctic Circle, the Sun does not rise at least once a year.
1 Arctic Circle: 66°30' N
2 Tropic of Cancer: 23°27' N
3 Equator: 0°
4 Tropic of Capricorn: 23°27' S
5 Antarctic Circle: 66°30' S

Map projections

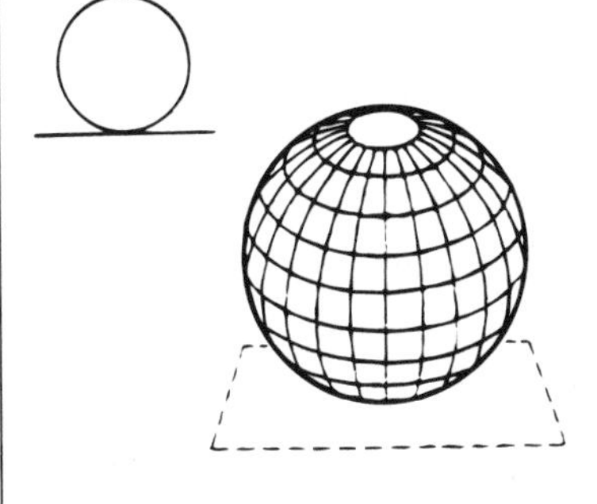

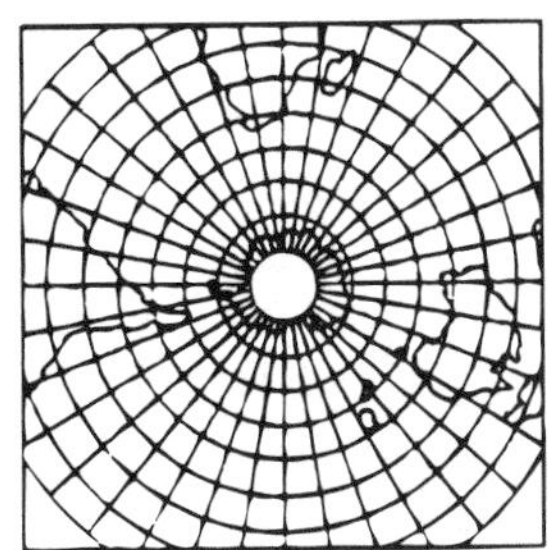

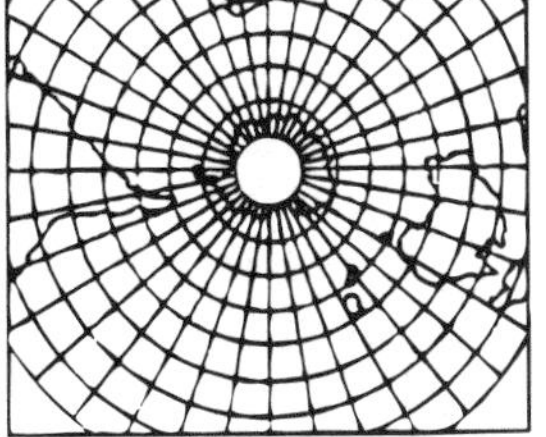

These are standard devices for showing the Earth's global surface on a flat sheet of paper. Each projection distorts the Earth in some way.

Azimuthal (zenithal) projection
This projects the Earth as if a flat sheet is touching the globe at the map center. It shows the shortest straight-line distances.

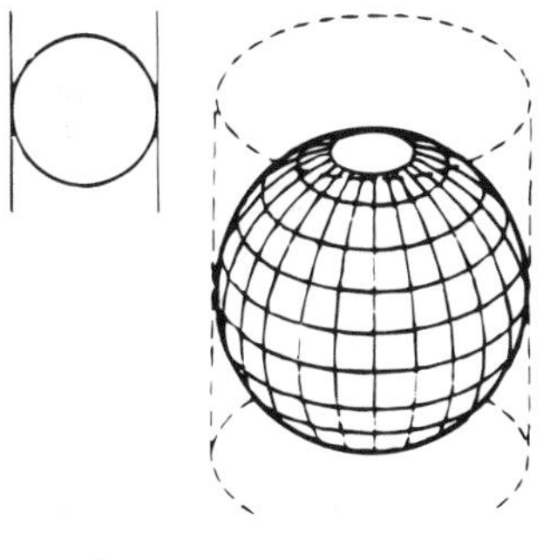

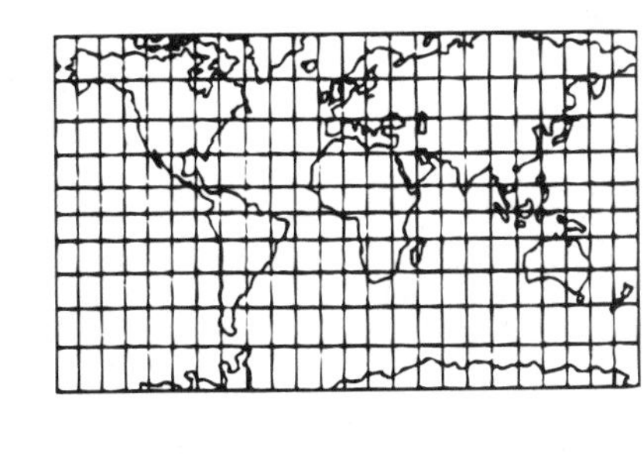

Cylindrical projection
This is made as if wrapping a sheet of paper around a globe's equator to produce a cylinder or tube. On such projections, lines of longitude meet lines of latitude at right angles and so do not meet at the poles. The view stretches polar areas but can show a true compass course.

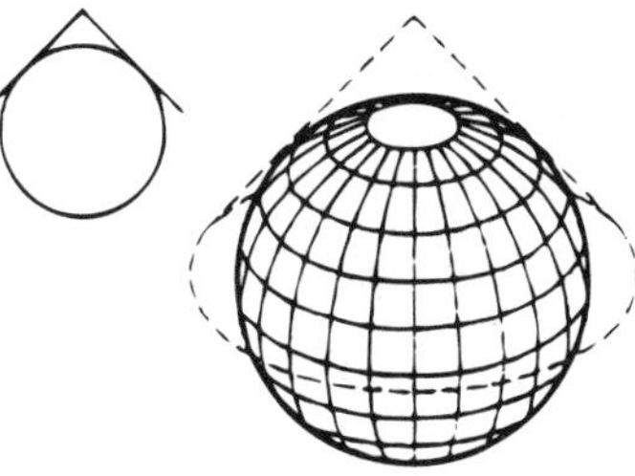

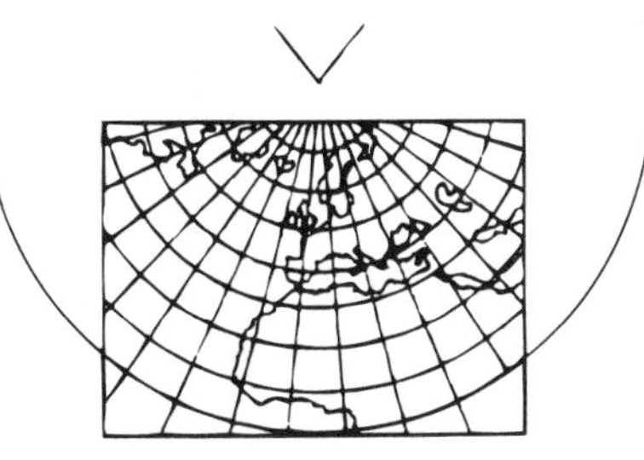

Conic projection
This is made as if a cone of paper is wrapped around a globe so as to touch it along one line of latitude. A conic projection shows lines of latitude as curved and lines of longitude as meeting at a pole. Conic projections show areas, directions, and distances fairly accurately.

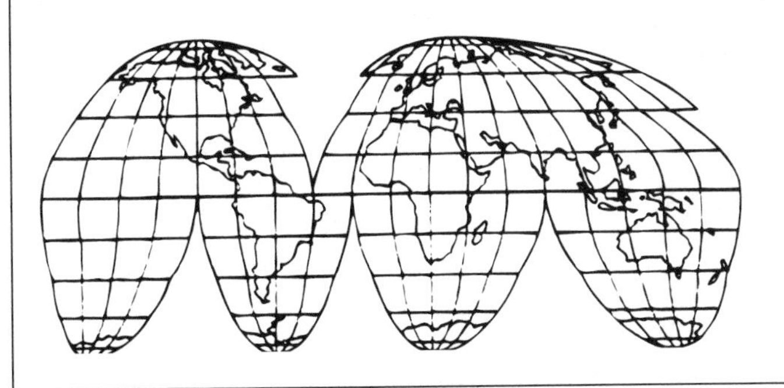

Mathematical projection
Mathematical projections are devised for special purposes. This homolosine equal-area projection is useful for showing the global distributions of different phenomena. Achieving accurate representation of area in this type of projection may involve interruption, as shown here.

Population structure

The population pyramid for Equatorial Guinea (1), a developing country, has a wide base (many births), reducing steadily to a narrow point (few old people). In contrast, the pyramid for France (2), a developed country, has a narrower base and a wide middle, denoting a low birth rate and many middle-aged people.

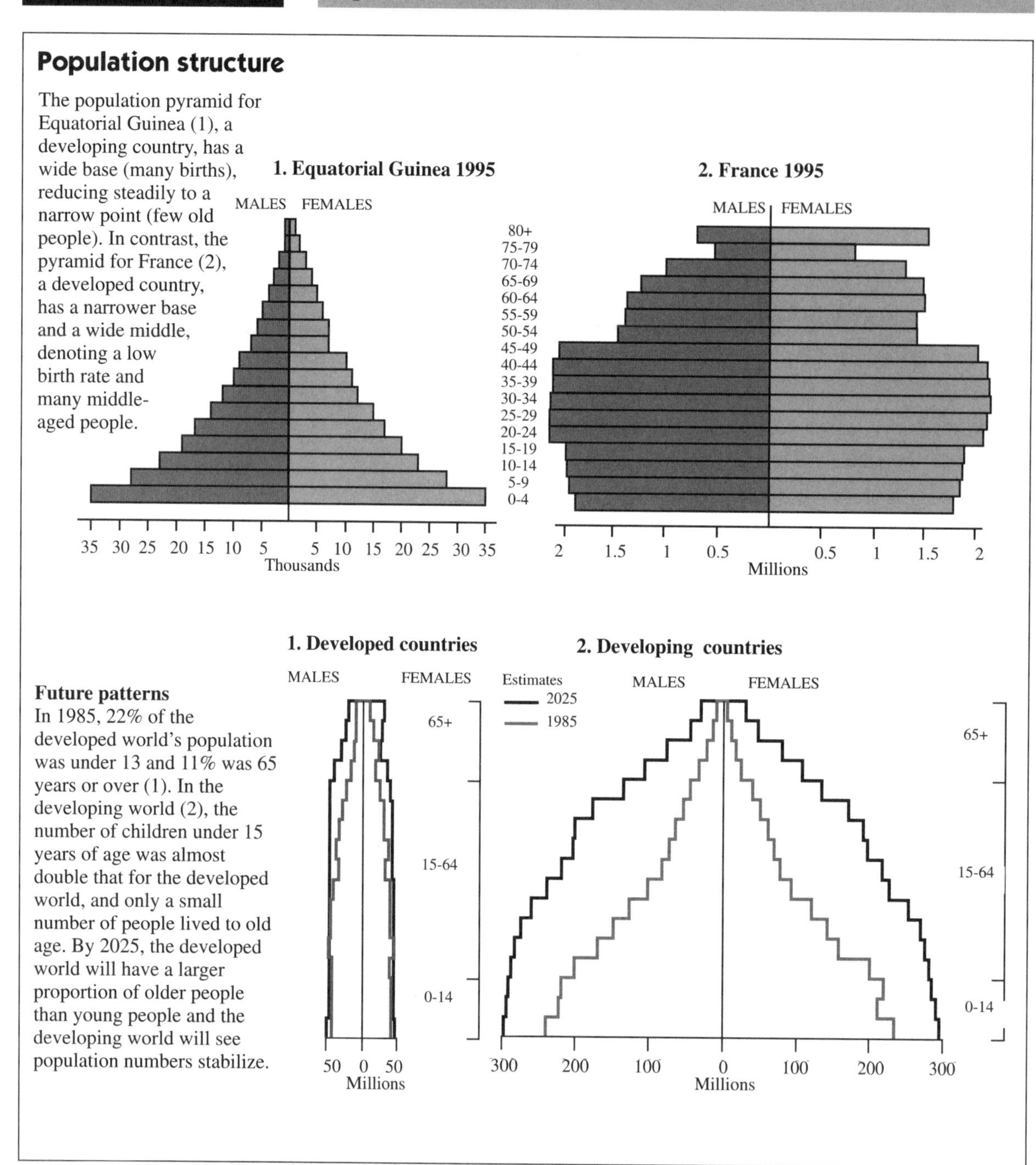

Future patterns
In 1985, 22% of the developed world's population was under 13 and 11% was 65 years or over (1). In the developing world (2), the number of children under 15 years of age was almost double that for the developed world, and only a small number of people lived to old age. By 2025, the developed world will have a larger proportion of older people than young people and the developing world will see population numbers stabilize.

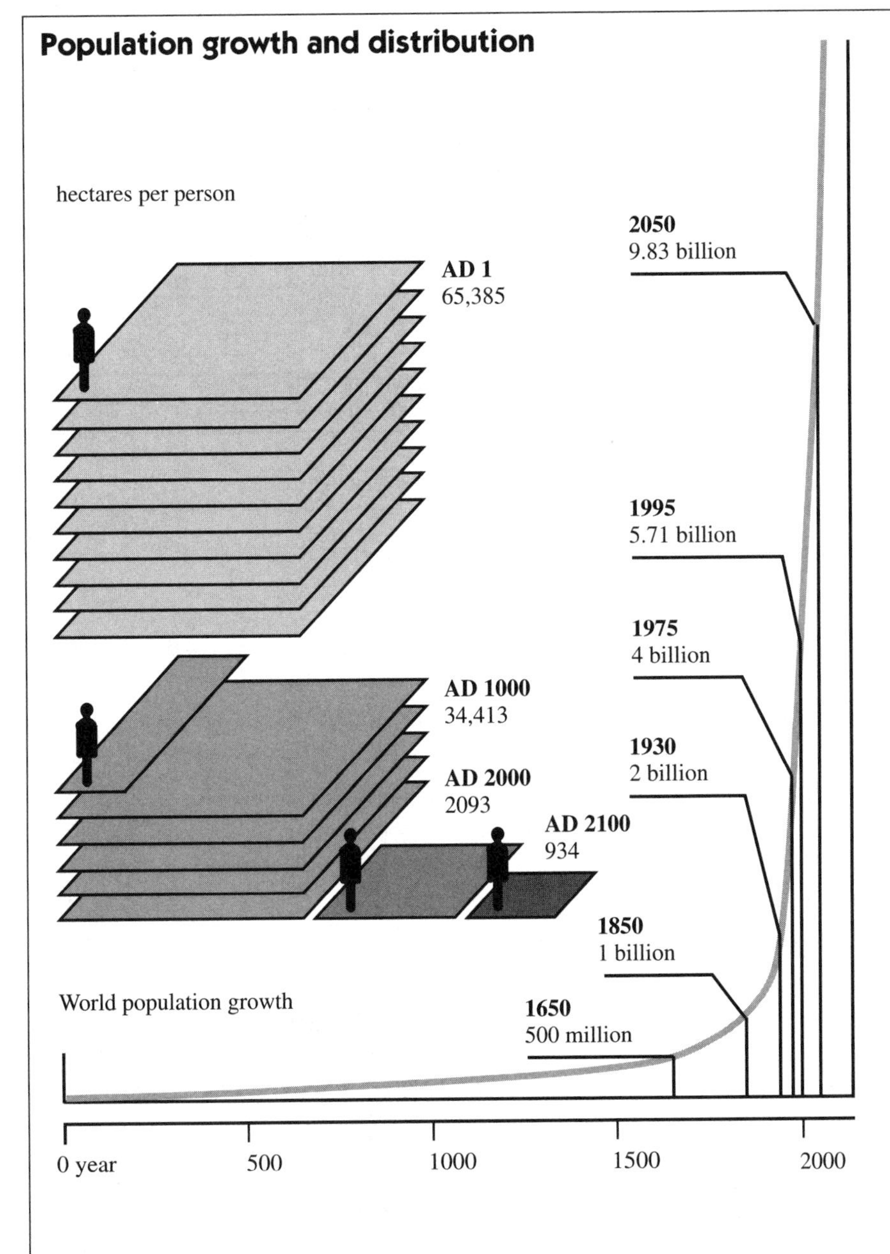

Over the past 2,000 years the population of the world has increased thirtyfold. At the time of Christ there were over 65,385 hectares of space per person. The rapid increase in world population over the past 150 years has been due to improvements in medical care, food resources and production, longer life expectancy, and a lower infant mortality rate.

Climatic changes

The greenhouse effect
Heat from the Sun (**1**) enters the atmosphere and heats the Earth (**2**). The heat is reflected by the Earth's surface (**3**) and some heat escapes into space (**4**). In large quantities, certain gases, such as carbon dioxide, build up in the atmosphere and trap some of the heat (**5**). If this happened to a greater degree, some of the polar ice caps would melt and sea levels would rise, flooding some land. Scientists do not know if these temperature rises will happen. Limits are now being placed on carbon dioxide emissions. In the mid-21st century, temperatures could rise as shown above right.

The ozone layer
The ozone layer (**6**) contains ozone, a form of oxygen, in the stratospheric layer of the Earth's atmosphere. It filters out the Sun's harmful ultraviolet rays (**7**), which can cause skin cancer. At present the ozone layer is being depleted as a result of chemical reactions started by chlorofluorocarbons (CFCs). Since the mid-1980s the use of CFCs in aerosols and refrigerators has been reduced. In spite of this, the ozone hole over Antarctica reached the size of North America in 1993.

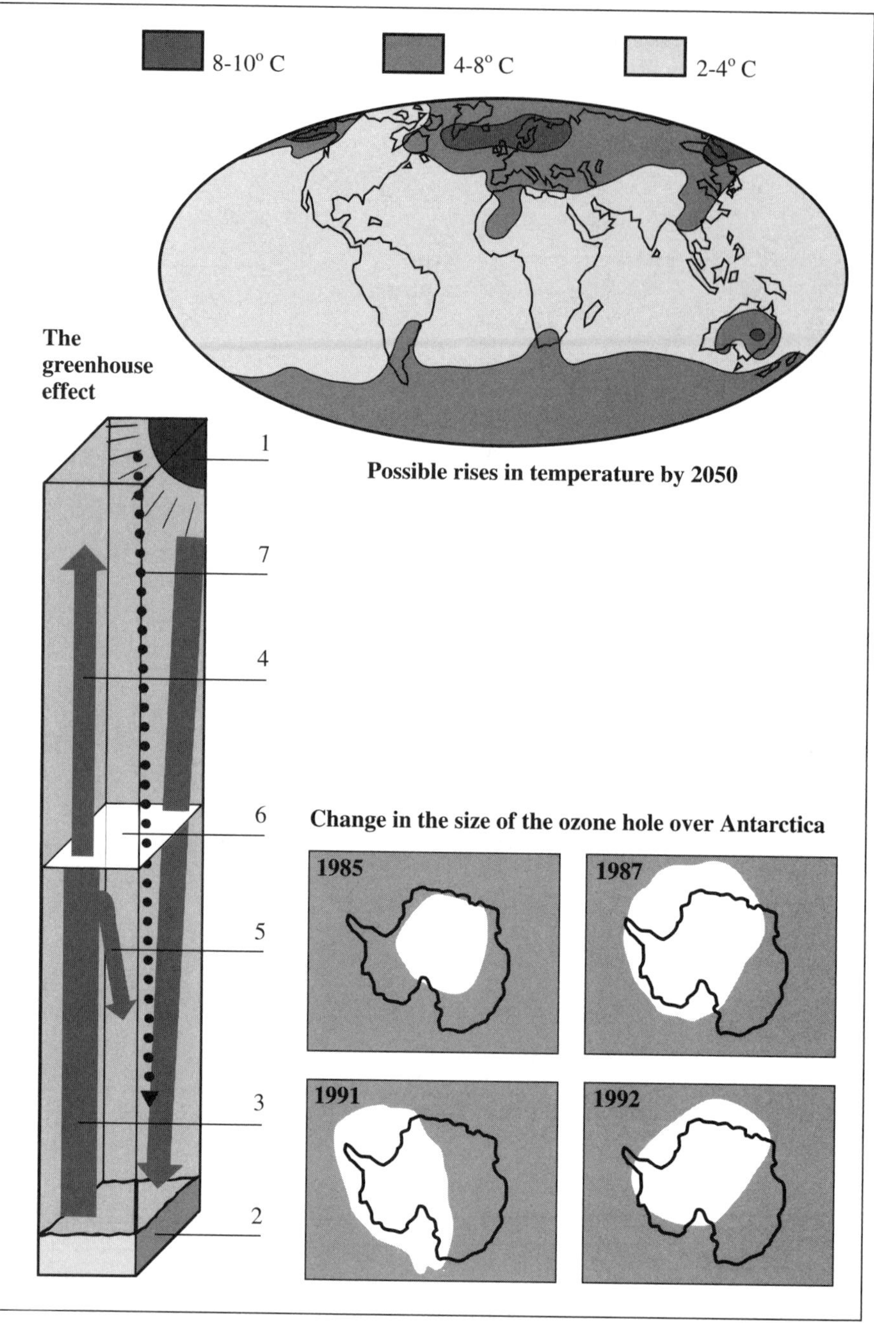

Possible rises in temperature by 2050

Change in the size of the ozone hole over Antarctica

Deforestation

Tropical forests destroyed since 1940

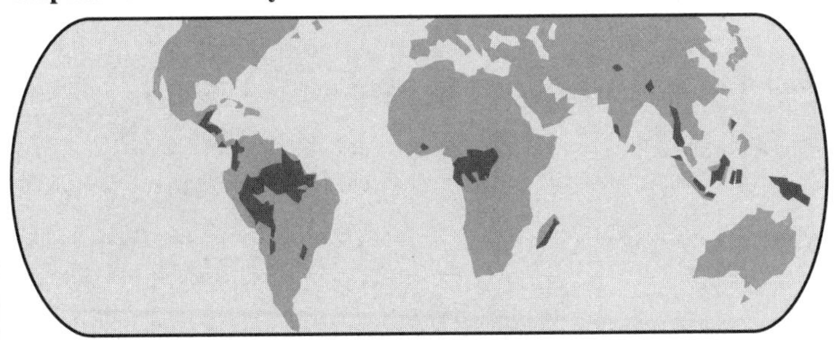

Tropical forests under threat in 1990

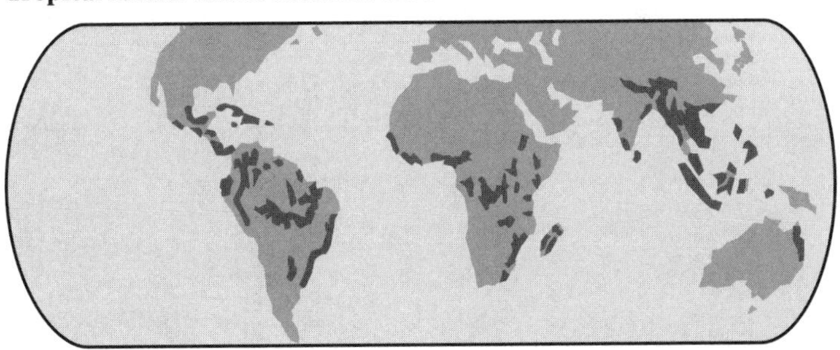

Since 1940 the vast tropical rainforests around the world have been greatly reduced in size. Large areas of forest have been destroyed. Most of the trees have been cut for timber and to clear land for farming and ranching.

Desertification

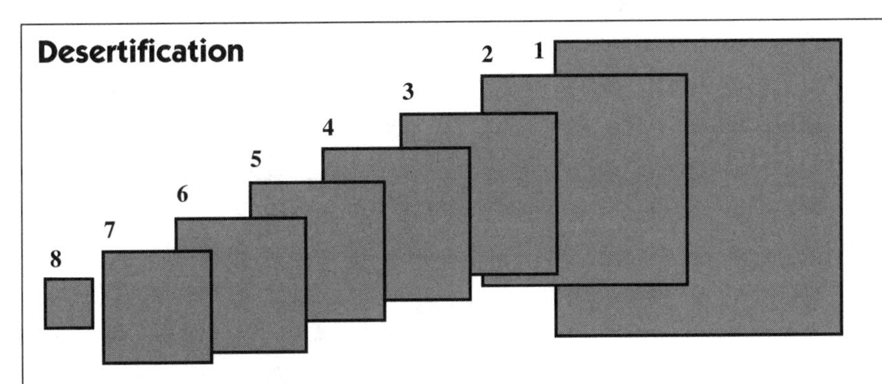

Deforestation, overgrazing, and intensive farming cause desertification. If the soil is dry, it is left exposed to the wind and rain, which strip away the topsoil. Agricultural lands that were once productive become infertile and lose their vegetation. The desert then extends and takes over the land.

Areas of land threatened by desertification

Figures in million hectares

1	Africa	739	**5**	Former USSR (in Asia)	164
2	Asia	365	**6**	North America	163
3	China and Mongolia	216	**7**	Australia	111
4	South America and Mexico	206	**8**	Mediterranean Europe	22

The planets

Distances from Sun

	Million km	Million mi	au
Pluto	5891	3661	39.36
Neptune	4497	2796	30.06
Uranus	2869	1784	19.18
Saturn	1427	887	9.54
Jupiter	778	484	5.20
Mars	227	141	1.52
Earth	150	93	1.00
Venus	108	67	0.72
Mercury	8	36	0.39

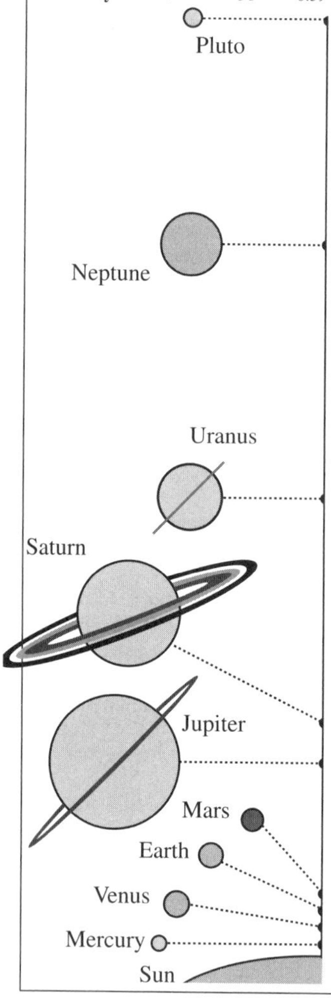

Pluto
Neptune
Uranus
Saturn
Jupiter
Mars
Earth
Venus
Mercury
Sun

Data	Mercury	Venus	Earth	Mars	Jupiter
Average distance from Sun	0.39 au	0.72 au	1.00 au	1.52 au	5.20 au
Distance at perihelion	0.31 au	0.72 au	0.98 au	1.38 au	4.95 au
Distance at aphelion	0.47 au	0.73 au	1.02 au	1.67 au	5.46 au
Closest distance to Earth	0.54 au	0.27 au		0.38 au	3.95 au
Average orbital speed	29.76 mi/sec (47.9 km/sec)	21.75 mi/sec (35.0 km/sec)	18.52 mi/sec (29.8 km/sec)	14.98 mi/sec (24.1 km/sec)	8.14 mi/sec (13.1 km/sec)
Rotation period	58 days 15 hr	243 days	23 hr 56 min	24 hr 37 min	9 hr 50 min
Sidereal period	88 days	224.7 days	365.3 days	687 days	11.86 years
Diameter at equator	3,031 mi (4,878 km)	7,521 mi (12,104 km)	7,926 mi (12,756 km)	4,222 mi (6,795 km)	88,731 mi (142,800 km)
Mass (Earth's mass = 1)	0.06	0.82	1	0.11	317.9
Surface temperature	333°F (167°C)	855°F (457°C)	59°F (15°C)	–125 to 23°F (–87 to –5°C)	–163°F (–108°C)
Surface gravity (Earth's gravity = 1)	0.38	0.88	1	0.38	2.64
Density (density of water = 1)	5.5	5.25	5.52	3.94	1.33
Number of satellites known	0	0	1	2	16
Number of rings known	0	0	0	0	1
Main gases in atmosphere	No atmosphere	Carbon dioxide	Nitrogen oxide	Carbon dioxide	Hydrogen/helium

Saturn	Uranus	Neptune	Pluto	Sun	Moon	
9.54 au	19.18 au	30.06 au	39.36 au		1.00 au	Average distance from Sun
9.01 au	18.28 au	29.80 au	29.58 au			Distance at perihelion
10.07 au	20.09 au	30.32 au	49.14 au			Distance at aphelion
8.00 au	17.28 au	28.80 au	28.72 au	0.98 au	0.0024 au	Closest distance to Earth
5.98 mi/sec (9.6 km/sec)	4.23 mi/sec (6.8 km/sec)	3.36 mi/sec (5.4 km/sec)	2.92 mi/sec (4.7 km/sec)		0.621 mi/sec (1 km/sec)	Average orbital speed
10 hr 14 min	16 hr 10 min	18 hr 26 min	6 days 9 hr	1 month	27 days 7 hr 43 min	Rotation period
29.46 years	84.01 years	164.8 years	247.7 years			Sidereal period
74,564 mi (120,000 km)	31,566 mi (50,800 km)	30,137 mi (48,500 km)	3,725 mi (5,995 km)	863,746 mi (1.39 mill km)	2,160 mi (3,476 km)	Diameter at equator
95.1	14.6	17.2	0.002–0.003	333,000	0.012	Mass (Earth's mass = 1)
–218°F (–139°C)	–323°F (–197°C)	–328°F (–200°C)	–355°F (–215°C)	99,000°F (55,000°C)	225–307°F (107–153°C)	Surface temperature
1.15	1.17	1.2	not known		0.165	Surface gravity (Earth's gravity = 1)
0.71	1.2	1.67	not known	0.1–100	3.34	Density (density of water = 1)
19	5	3	1	9 planets	0	Number of satellites known
1,000-	9	0	0	0	0	Number of rings known
Hydrogen	Hydrogen/helium, methane	Hydrogen/helium, methane	Methane	Hydrogen/helium		Main gases in atmosphere

The planets

CHARTS & TABLES

The phases of the Moon, rotations, and tides

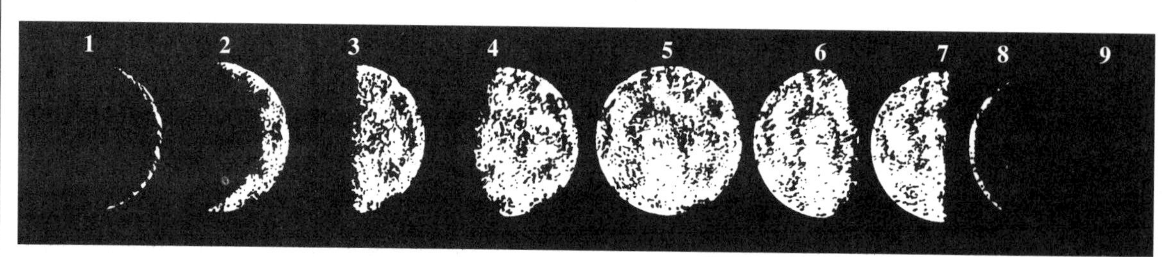

The Moon produces no light of its own: it shines because it reflects sunlight. The amount of the lit half that can be seen from the Earth changes from day to day. These regular changes are known as phases of the Moon. Here we show how the Moon appears from the Earth at different stages during the month. The interval between one new Moon and the next is 29 days, 12 hr, 44 min, 3 sec.

Names of the phases
A waxing Moon is one that becomes increasingly visible; a waning Moon becomes less visible; a gibbous ("humped") Moon is between the half and full phases.

1 New Moon
2 Waxing crescent Moon
3 Half Moon, first quarter
4 Waxing gibbous Moon
5 Full Moon
6 Waning gibbous Moon
7 Half Moon, last quarter
8 Waning crescent Moon
9 New Moon

Rotations Here is shown the relationship between the Earth, the Sun, and the Moon. Arrows indicate the direction of rotation.

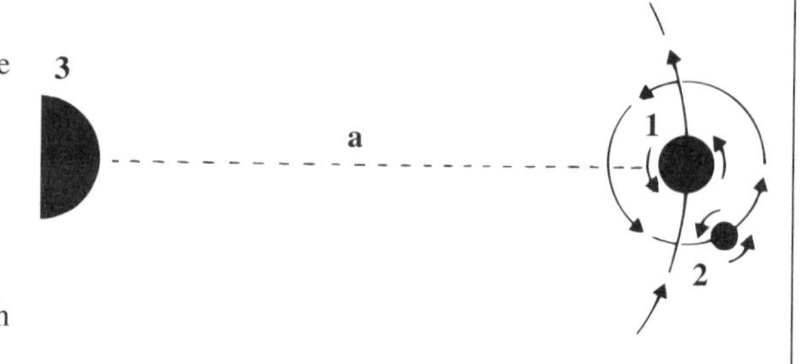

1 The Earth The Earth's mean distance from the Sun is 93 million mi (150 million km) (**a**). The Earth orbits the Sun in 365.25 days. Meanwhile, the Earth rotates every 23 hr 56 min.

2 The Moon It takes 27.3 days for the Moon to complete one orbit of the Earth. Because it takes exactly the same time for the Moon to turn once on its own axis, the same side of the Moon always faces the Earth. The far side of the Moon can be seen only from space.

3 The Sun The Sun, like the Earth, rotates on its axis. Because the Sun is made of gas, it can rotate at different speeds at different latitudes: it rotates more slowly at its poles than at its equator.

Tides The pull of the Moon's gravity is the main cause of the tides in the Earth's seas and oceans, although the Sun also has an influence. The highest ("spring") tides occur when the Moon and Sun pull along the same line (1 and 2); the lowest ("neap") tides occur when they pull at right angles (3 and 4).

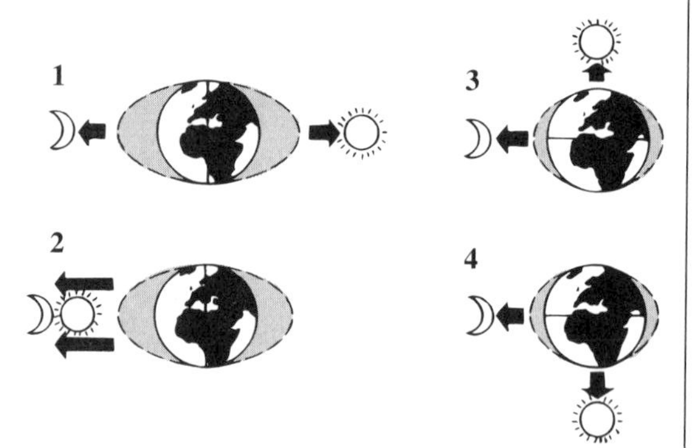

Eclipses

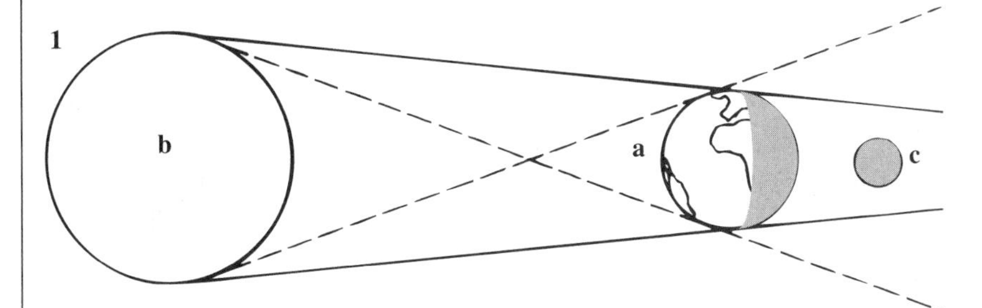

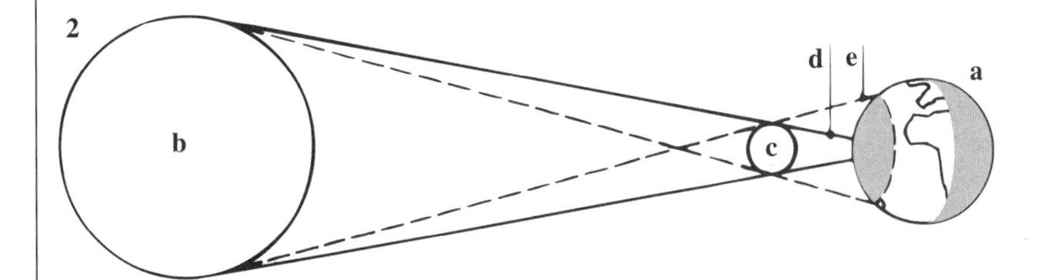

1 Eclipse of the Moon This occurs when the Earth (a) is in a direct line between the Sun (b) and Moon (c). The Moon is then in the Earth's shadow and cannot receive any direct sunlight. It becomes dim and appears coppery red in color. There are never more than three eclipses of the Moon in a year.

2 Eclipse of the Sun From the Earth, the Sun and Moon appear to be about the same size: the Sun is about 400 times as big as the Moon but is also 400 times farther away from the Earth. When the Moon (c) is in a direct line between the Sun (b) and the Earth (a), the Moon's disk-shaped outline appears to cover the Sun's bright surface, or photosphere. The part of the Earth directly in the Moon's shadow (d) sees a total eclipse of the Sun; areas around it (e) see a partial eclipse.

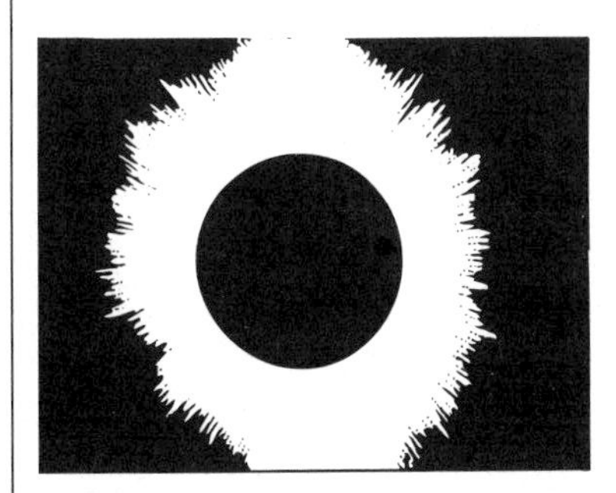

Total This can last from a split second up to a maximum of 7 min 31 sec. The area over which it is seen may have a maximum width of 169 miles (272 km), but is usually much less. The corona – the circle of light that appears as a halo around the Moon's disk during this type of eclipse – is visible to the naked eye.

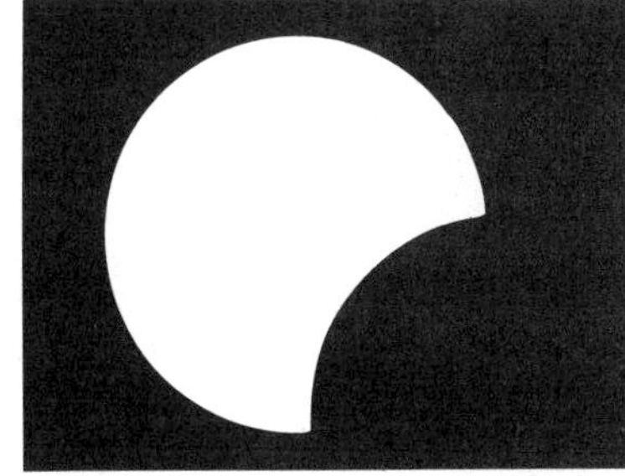

Partial In this type of eclipse, the Moon's disk obscures only part of the photosphere.

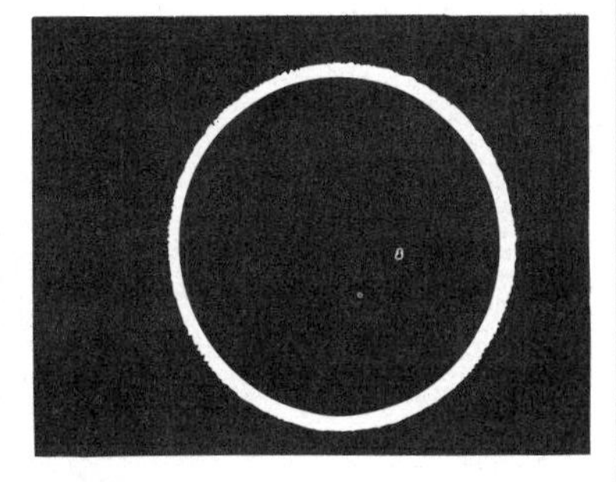

Annular This type of eclipse, named from the Latin word *annulus*, meaning "ring," occurs if the Moon is at its farthest point from the Earth and the Earth is at its nearest point to the Sun. The Moon's disk then appears slightly smaller than the photosphere.

The universe

Astronomers estimate that the universe is 13–20 billion years old. It is expanding as the galaxies and clusters of galaxies move further away from each other.

The cubes, right, represent the enormous scale of the universe. Each cube has sides 100 times as long as the sides of the cube before it. The length of the sides is in light-years (ly).

1 Cube side 950 au (0.015 ly). Contains the whole solar system.

2 Cube side 1.5 ly. Contains the solar system surrounded by the Oort cloud of comets. This cloud may be the source of many comets that pass through the solar system. It surrounds the Sun at an average distance of 40,000 au (2/3 ly).

3 Cube side 150 ly. Contains the solar system and the nearer stars.

4 Cube side 15,000 ly. Contains the nearer spiral arms of our galaxy.

5 Cube side 1.5 million ly. Contains the whole of our galaxy, the large and small Magellanic Clouds, and other nearby galaxies in the local group.

6 Cube side 150 million ly. Contains the whole of the local group and the Pisces, Cancer, and Virgo clusters of galaxies.

7 Cube side 15 billion ly Contains all the known clusters and superclusters of galaxies and all other known objects in space.

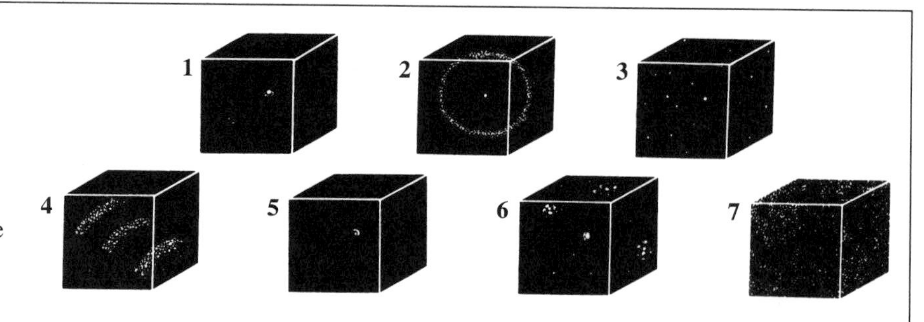

Units of space measurement

Astronomical unit (au)

Mean Sun to Earth distance = 92,955,807 miles (149,597,870 km). Agreed internationally in 1964 but value has altered.

Galactic coordinates

Relative location of our galaxy's components in latitude and longitude (degrees and min) measured in relation to the celestial equator, which is a projection of the Earth's equator.

Light-year (ly)

Distance traveled by light in a year = 5.878 trillion mi (9.4605 trillion km) or 63,290 au. Defined in 1888.

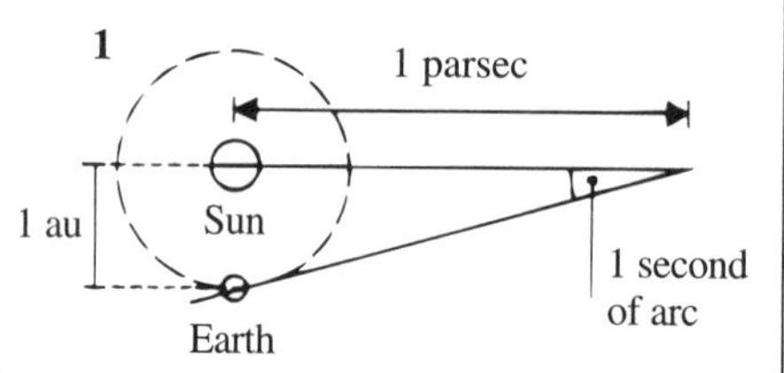

Parsec (pc)

The distance at which 1 au would measure 1 sec of arc = 19.16 trillion mi (30.857 trillion km) or 206,265 au or 3.26 ly (1).

Astronomical time

Time can be measured by motion; in fact, the motion of the Earth, Sun, Moon, and stars provided humans with the first means of measuring time.

Years, months, days

Sidereal times are calculated by the Earth's position according to fixed stars. The anomalistic year is measured according to the Earth's orbit in relation to the perihelion (Earth's minimum distance to the Sun). Tropical times refer to the apparent passage of the Sun and the actual passage of the Moon across the Earth's equatorial plane. The synodic month is based on the phases of the Moon. Solar time (as in a mean solar day) refers to periods of darkness and light averaged over a year.

Time	Days	Hours	Minutes	Seconds
sidereal year	365	6	9	10
anomalistic year	365	6	13	53
tropicalyear	365	5	48	45
sidereal month	27	7	43	11
tropical month	27	7	43	5
synodic month	29	12	44	3
mean solarday	0	24	0	0
sidereal day	0	23	56	4

Galaxies

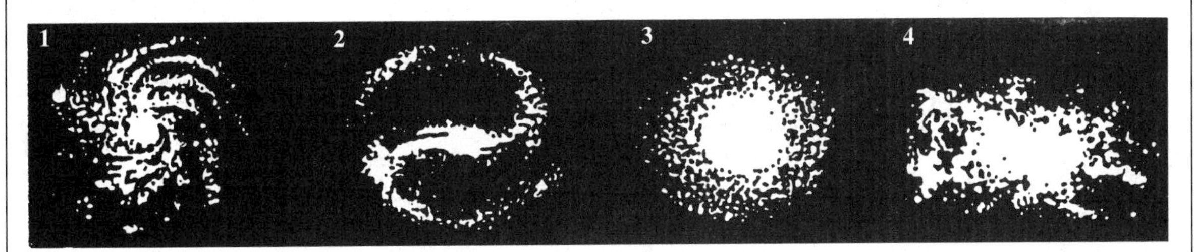

Galaxies are collections of stars and planets and clouds of gas or dust that form "islands" in the emptiness of space. A recent theory claims much of this is occupied by invisible dark matter. Most galaxies are found in groups; very few are found on their own.

Galactic shapes
Galaxies are classified by their shape. The four main classes are:

1 Spiral These galaxies resemble pinwheels, with spiral arms trailing out from a bright center. Our galaxy is a spiral galaxy at the center of which is a cluster of stars called the Milky Way.
2 Barred spiral Here, the spiral arms trail from the ends of a central bar. About 30% of galaxies are spirals or barred spirals.
3 Elliptical These galaxies do not have spiral arms. About 60% of galaxies are elliptical, varying in shape from almost spherical (like a soccer ball) to very flattened (like a football).
4 Irregular About 10% of galaxies are irregular with no definite shape.

Life cycle of a star

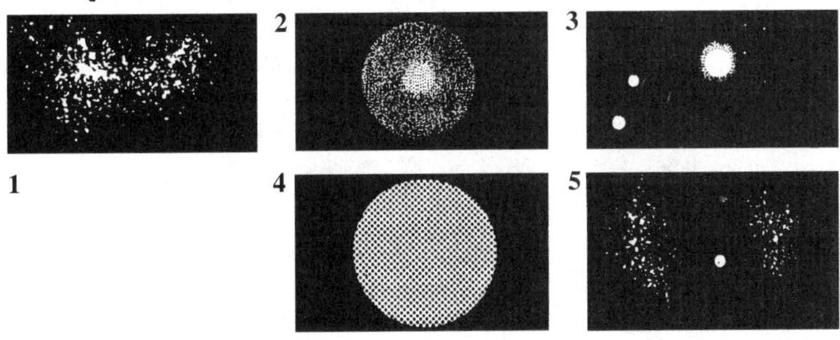

The life cycle of a star with a mass similar to that of the Sun:
1 All stars form in nebulae (clouds of gas and dust). Denser, smaller clouds called globules gradually form within a nebula. (A globule the size of the solar system will form a star the size of the Sun.)
2 The globule becomes smaller and hotter, begins to shine faintly, and forms a large, red "protostar" (star ancestor). "Protoplanets" may form around the protostar.
3 The protostar contracts further, gets hotter, and becomes an ordinary star (one producing energy by converting hydrogen to helium). The star and any planets remain stable for a few billion years. (This is the current state of the Sun.)
4 Eventually, the star's core gets hotter while its outer layers expand and become cooler and redder. The star becomes a red giant, destroying any planets close to it. (The Sun is expected to become a red giant, 100 times its present size, in about 5 billion years. It will engulf Mercury, Venus, and possibly the Earth.)
5 When the red giant reaches its maximum size, it becomes unstable and pulsates (swells and shrinks). The outer layers break away to form a ring nebula. The core shrinks to form a white dwarf star.

Length and Area

The first group of tables converts imperial to metric and metric to imperial. The second group converts imperial to metric.

Milli-inches to Micrometers

mils	µm
1	25.4
2	50.8
3	76.2
4	101.6
5	127.0
6	152.4
7	177.8
8	203.2
9	228.6
10	254.0
20	508.0
30	762.0
40	1016.0
50	1270.0
60	1524.0
70	1778.0
80	2032.0
90	2286.0
100	2540.0

Inches to Millimeters

in	mm
1	25.4
2	50.8
3	76.2
4	101.6
5	127.0
6	152.4
7	177.8
8	203.2
9	228.6
10	254.0
20	508.0
30	762.0
40	1016.0
50	1270.0
60	1524.0
70	1778.0
80	2032.0
90	2286.0
100	2540.0

Inches to Centimeters

in	cm
1	2.54
2	5.08
3	7.62
4	10.16
5	12.70
6	15.24
7	17.78
8	20.32
9	22.86
10	25.40
20	50.80
30	76.20
40	101.60
50	127.00
60	152.40
70	177.80
80	203.20
90	228.60
100	254.00

Feet to Meters

ft	m
1	0.305
2	0.610
3	0.914
4	1.219
5	1.524
6	1.829
7	2.134
8	2.438
9	2.743
10	3.048
20	6.096
30	9.144
40	12.192
50	15.240
60	18.288
70	21.336
80	24.384
90	27.432
100	30.480

Yards to Meters

yd	m
1	0.914
2	1.829
3	2.743
4	3.658
5	4.572
6	5.486
7	6.401
8	7.315
9	8.230
10	9.144
20	18.288
30	27.432
40	36.576
50	45.720
60	54.864
70	64.008
80	73.152
90	82.296
100	91.440

Fathoms to Meters

fm	m
1	1.83
2	3.66
3	5.49
4	7.32
5	9.14
6	10.97
7	12.80
8	14.63
9	16.46
10	18.29
20	36.58
30	54.87
40	73.16
50	91.45
60	109.74
70	128.03
80	146.32
90	164.61
100	182.90

Circular mils to Square micrometers

cmil	µm²
1	506.7
2	1013.4
3	1520.1
4	2026.8
5	2533.5
6	3040.2
7	3546.9
8	4053.6
9	4560.3
10	5067.0
20	10 134.0
30	15 201.0
40	20 268.0
50	25 335.0
60	30 402.0
70	35 469.0
80	40 536.0
90	45 603.0
100	50 670.0

Square inches to Square millimeters

in²	mm²
1	645.2
2	1290.4
3	1935.6
4	2580.8
5	3226.0
6	3871.2
7	4516.4
8	5161.6
9	5806.8
10	6452.0
20	12 904.0
30	19 356.0
40	25 808.0
50	32 260.0
60	38 712.0
70	45 164.0
80	51 616.0
90	58 068.0
100	64 520.0

Square inches to Square centimeters

in²	cm²
1	6.452
2	12.903
3	19.355
4	25.806
5	32.258
6	38.710
7	45.161
8	51.613
9	58.064
10	64.516
20	129.032
30	193.548
40	258.064
50	322.580
60	387.096
70	451.612
80	516.128
90	580.644
100	645.160

Square feet to Square meters

ft²	m²
1	0.093
2	0.186
3	0.279
4	0.372
5	0.465
6	0.557
7	0.650
8	0.743
9	0.836
10	0.929
20	1.858
30	2.787
40	3.716
50	4.645
60	5.574
70	6.503
80	7.432
90	8.361
100	9.290

Temperature

Systems of measurement

Below, the different systems of temperature measurement are compared: Fahrenheit (°F), Celsius (°C), Réaumur (°r), Rankine (°R) and Kelvin (K).

a) boiling point of water **c**) absolute zero

b) freezing point of water

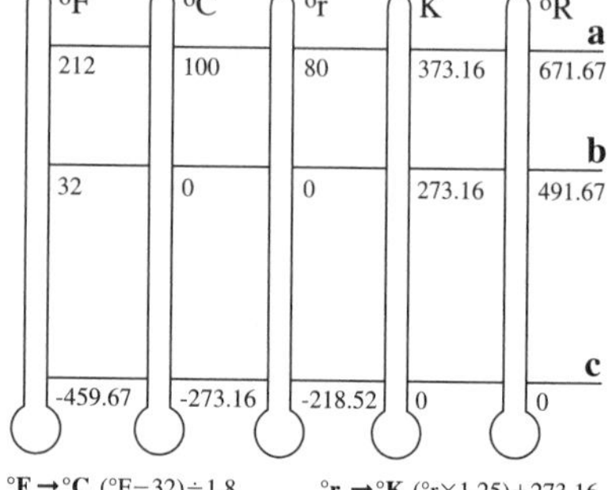

°F → °C (°F−32)÷1.8

°C → °F (°C×1.8)+32

°F → °K (°F+459.67)÷1.8

°C → °K °C+273.16

°r → °K (°r×1.25)+273.16

°R → °K °R÷1.8

°K → °F (°K×1.8)−459.67

°K → °C °K−273.16